New Horizons
An AOCS/CSMA Detergent Industry Conference

New Horizons
An AOCS/CSMA Detergent Industry Conference

Editor

Richard T. Coffey

FMC Corp.
Princeton, New Jersey

Champaign, Illinois

AOCS Mission Statement

To be a forum for the exchange of ideas, information, and experience among those with a professional interest in the science and technology of fats, oils, and related substances in ways that promote personal excellence and provide high standards of quality.

AOCS Books and Special Publications Committee

E. Perkins, chairperson, University of Illinois, Urbana, Illinois
N.A.M. Eskin, University of Manitoba, Winnipeg, Manitoba
M. Pulliam, C&T Quincy Foods, Quincy, Illinois
T. Foglia, USDA—ERRC, Wyndmoor, Pennsylvania
L. Johnson, Iowa State University, Ames, Iowa
Y.-S. Huang, Ross Laboratories, Columbus, Ohio
J. Lynn, Lever Brothers Co., Edgewater, New Jersey
M. Mossoba, Food and Drug Administration, Washington, D.C.
G. Nelson, Western Regional Research Center, San Francisco, California
F. Orthoefer, Stuttgart, Arkansas
J. Rattray, University of Guelph, Guelph, Ontario
A. Sinclair, Royal Melbourne Institute of Technology, Melbourne, Australia
G. Szajer, Akzo Chemicals, Dobbs Ferry, New York
B. Szuhaj, Central Soya Co., Inc., Fort Wayne, Indiana
L. Witting, State College, Pennsylvania

Copyright © 1996 by AOCS Press. All rights reserved. No part of this book may be reproduced or transmitted in any form or by any means without written permission of the publisher.

The paper used in this book is acid-free and falls within the guidelines established to ensure permanence and durability.

Library of Congress Cataloging-in-Publication Data

New horizons : an AOCS/CSMA detergent industry conference / editor,
  Richard T. Coffey
      p. cm.
   Includes bibliographical references and index.
   ISBN 0-935315-78-0 (alk. paper)
   1. Detergents--Congresses. I. Coffey, Richard T.
TP992.5.N49 1996
668′.1--dc21   96-49995
                                                                        CIP

Printed in the United States of America with vegetable oil-based inks.
00 99 98 97 96 5 4 3 2 1

Preface

In September 1995 in Lake George, New York, two of the leading organizations in the field of surfactants and detergents [the American Oil Chemists' Society (AOCS) and the Chemical Specialties Manufacturers' Association (CSMA)] combined to present the third industrywide conference to inform and update attendees on various aspects of the detergents and cleaning products industry. The meeting, titled New Horizons '95, offered a technological update on the consumer and industrial and institutional detergents industry in the United States and abroad.

Topics covered at the conference, and presented in the collection of papers which follow, include new surfactants materials, new uses for current raw materials, new developments in the technology of formulations, and the needs and outlook of the detergents industry. This publication should be a valuable tool for a wide range of professionals in the surfactants and detergents industry and will offer readers the opportunity to better understand current issues in our industry, both technically and in the marketplace.

Ken Schoene
Conference Chairperson

Terri Germain
Conference Co-chairperson

Contents

Preface v

Chapter 1 Safety and Regulatory Issues Affecting Cleaning Products 1
           Richard Sedlak

Chapter 2 Surfactant Challenges for 2000 and Beyond 18
           David J. Kitko

Chapter 3 A New and Promising Chapter in
           the Sodium Silicate Story 23
           Hans-Peter Rieck

Chapter 4 Advances in Detergent Processing 37
           Michael Hill and Robert Ahart

Chapter 5 Polymers for Detergents: Current Technology
           and Future Trends 42
           William L. McCullen

Chapter 6 Detergent Enzymes—Global Trends 57
           Hans A. Hagen

Chapter 7 On Understanding Hydrophobicity 63
           Bengt Kronberg, Miguel Costas, and Rebecca Silveston

Chapter 8 Gemini Surfactants 70
           Milton J. Rosen

Chapter 9 Surfactants in the Environment 79
           John F. Scamehorn, Sherril D. Christian,
           Jeffrey H. Harwell, and David A. Sabatini

Chapter 10 New Chelating Agents for the Detergent
           and Cleaning Industry 97
           Werner W. Bertleff

Chapter 11 Mechanism of Enzyme Action and Correlation
           with Cleaning Performance 113
           Peter F. Plank, Stephen J. Danko, Judy Dauberman,
           Matthew J. Flynn, Constance Hsia,
           Deborah S. Winetzky, and Edward D. Cesare

Chapter 12 The Cost of New Product Development
           in Chemical Specialties 126
           Daniel R. Dutton

Chapter 13 Specialty Chemicals for Washing PET Beverage Bottles 130
           James Lichorat

Chapter 14	Eco-Efficiency: Industry's Path to Sustainability .... 134 *Kenneth Alston*	
Chapter 15	The Need for Multifunctional Surfactants .... 139 *Arshad Malik, Ned Rockwell, and Y.K. Rao*	
Chapter 16	Overview of European Regulatory Activities .... 149 *Pierre V. Costa*	
Chapter 17	Generation 2000 Appliances .... 154 *John T. Weizeorick*	
	Index 159	

Chapter 1

Safety and Regulatory Issues Affecting Cleaning Products

Richard Sedlak

The Soap and Detergent Association, 475 Park Avenue South, New York, NY 10016

Introduction

The soap and detergent industry recognizes that the safety of cleaning products and their contributions to the health and well-being of our society are substantial and well documented. It is also recognized that the achievements of this industry are widely respected by the scientific community. Yet concerns continue to be raised by a variety of groups about the safety of cleaning products and their ingredients. Left unanswered, these concerns could threaten to erode the confidence of consumers concerning not only the safety of formulated products, but also their performance and value.

To address these concerns effectively, it is important to first understand some of the changes that have occurred in the arenas in which these issues are discussed:

- The universe of participants, or stakeholders, has grown. Groups with concerns about cleaning products now include not only legislative and regulatory bodies, public interest groups, and the media, but also such traditional allies as consumer educators, professional organizations, academicians, and end users. Further, on the industry side, raw material suppliers and distributors, as well as product formulators, are being asked to participate in programs addressing the issues.
- The number and range of products and ingredients that raise concerns have expanded. In the past, safety questions focused typically on a limited number of specific ingredients. Today, however, the challenges are much broader. At The Soap and Detergent Association (SDA), for example, we see the potential for all cleaning product ingredients to be subject to scrutiny. Further, concerns previously limited to industrial products and the workplace are now being raised about cleaning products and their use in the home.
- Activity at the state and local levels has increased as authority over management of these issues has become more decentralized; and despite the new "common sense" attitude in Washington, many regulatory programs continue to develop at the federal level as a result of existing laws.
- Pollution prevention programs are being established as government agencies at all levels attempt to achieve environmental goals with fewer funds. From the standpoint of cleaning products, these pollution prevention initiatives are often in the form of environmental specifications for products procured by government agencies or recommendations appearing in public education programs. Such efforts are typically developed under administrative procedures which have lower require-

ments for proving their value than do legislative and regulatory programs, thus reducing burdens for government agencies. This feature plus the fact that government agencies often have a higher level of credibility than do other stakeholders makes such initiatives troubling to the industry.

These key changes as well as concerns being raised about our products appear to be driven by two types of forces. The first type reflects the flawed science of industry critics as evidenced by: confusion over human *vs.* environmental safety; failure to apply scientifically based risk assessment procedures in safety decisions; failure to use life-cycle assessment appropriately; and a focus on hazard rather than risk.

The second type of driving force reflects consumer conditions and includes: skepticism about industry in general; chemophobia; the natural inclination to safeguard home and family; an eagerness to "do something" to protect the environment; misinformation from industry critics about formulated cleaning products; a willingness to accept seemingly simple solutions to complex issues; and a lack of understanding of product benefits.

Taken together, these forces are having two major impacts: proposals for legislation and regulation, as well as government-sponsored education and procurement programs that are not based on sound science; and a growing institutionalized prejudice against formulated products.

SDA's approach to addressing issues of concern to the industry has historically been and continues to be to develop and share scientifically sound information with key stakeholders, such as the technical community, government bodies, educators, communicators, and consumers. Current SDA programs cover a range of human and environmental safety issues and are aimed both at protecting the ability of industry members to formulate products that best meet consumer needs, and maintaining consumer confidence in the safety and benefits of cleaning products.

This presentation will first review several regulatory and safety issues on which there is current government activity and which illustrate the changes described above; it will then move on to a summary of some other current issues, before concluding with a look at several emerging issues for the cleaning products industry.

Current Regulatory and Safety Issues

Volatile Organic Compounds (VOC)

The VOC issue is a key example of an area in which research has enabled industry to respond to government initiatives and avert inappropriate regulation. The contribution of VOC to the formation of ground level ozone, which is a hazard to animals and plants, has led to restrictions on VOC emissions from mobile (e.g., automobile) and stationary (e.g., industrial facilities) sources as a primary means of reducing ozone. Nonetheless, tens of millions of people live in areas that still exceed the national standard for ozone. Because of the regional occurrence of ozone problems, VOC are the subject of regulation at both the federal and state levels. The issue has also fostered coordinated regulatory efforts among jurisdictions within regions that span across airsheds.

As a result of the continuing noncompliance with national air quality standards, nonpoint sources of VOC—including consumer and commercial products—have been targeted by regulators, even though they are relatively small contributors to VOC emissions. Cleaning products have come under scrutiny because they contain ingredients defined as VOC (e.g., ethanol, isopropanol). Although the VOC used in cleaning products are generally present at relatively low concentrations, many product categories represent high total volume use of VOC due to significant product consumption.

Many cleaning products are disposed down-the-drain where most of their VOC biodegrade and, thus, are not emitted. However, regulators have often begun with the premise that most, if not all of the VOC in cleaning products end up being emitted into the atmosphere. A number of years ago, SDA recognized that the development of data on cleaning products that focused on this issue would greatly assist regulators. SDA undertook a technical research program to determine the fate of detergent VOC disposed down-the-drain and, therefore, the impact of VOC used in cleaning products on the formation of ozone. The release of ethanol into air during conditions that simulated consumer laundering and hand dishwashing was measured. In a complementary study, the fate of ethanol in both sewers and municipal wastewater treatment plants was modeled. Based on these studies, it was determined that less than 5% of the ethanol in hand dishwashing detergents and less than 1% of the ethanol in laundry detergents were released into the atmosphere. The remainder is biodegraded.

These findings led the California Air Resources Board (ARB) to delete provisions in proposed regulations that would have imposed VOC content limits on laundry and hand dishwashing detergents. Because most states that have subsequently considered regulations on the VOC content of consumer products have used the existing California regulation as a starting point, there has been no serious consideration of regulating these detergent products. However, a number of categories of hard surface cleaning products that were regulated by California appear in the regulations of other states, including bathroom and tile cleaners, general purpose cleaners, glass cleaners, and oven cleaners.

Late in 1994, ARB proposed a revised State Implementation Plan (SIP) for ozone in an effort to avoid imposition of a Federal Implementation Plan. The SIP requires a reduction in emissions of VOC from consumer products by 85% from their 1990 levels by the year 2010. To comply with the SIP, ARB began a new round of regulation development in spring 1995 which it plans to complete in 1997.

ARB has issued a draft proposal categorizing product categories identified by EPA in relation to their relative emissions and the potential to achieve reductions as a result of regulations. Of major categories of interest to the soap and detergent industry, ARB classified several, such as machine dishwashing detergents, as having total emissions that are too low to consider significant. It also placed hand dishwashing and laundry detergents in the lowest priority of three classifications of product categories that should be evaluated for possible regulation.

SDA is participating in a multistakeholder committee and work groups formed by ARB to assist with the development of the regulations. The work groups are reviewing what is known about photoreactivity (i.e., the relative amount of ozone contributed to the air per unit mass of a compound) and other information needed by ARB to develop regulations. ARB intends to take the varying photoreactivity of VOC into account in the next round of regulations, and to modify existing consumer product VOC regulations to account for photoreactivity.

The most significant activity at the federal level related to consumer product VOC has been EPA's Report to Congress, which was submitted in spring 1994. Mandated by the 1990 amendments to the Clean Air Act, the EPA report included a national inventory of VOC emissions from consumer and commercial products. Based on the report, EPA has issued a notice of the categories of products that it intends to propose for regulation.

EPA based the inventory on a national survey of manufacturers conducted in 1993 to determine the VOC content and sales of consumer and commercial products. Manufacturers of cleaning products were included in the survey. To assist the agency, SDA submitted a detailed evaluation of the EPA survey results for laundry and dishwashing products disposed down-the-drain, along with estimates of VOC emissions from these products. EPA adopted SDA's emission estimates for these product categories.

In July 1994, EPA convened a standing advisory committee to assist in ranking 96 categories of consumer and commercial products for regulatory action based on market volumes, emissions, product functionality, and the potential for reformulation. The only laundry and dishwashing products included in the ranking exercise were laundry and hand dishwashing detergents, which were ranked 94th and 96th, respectively.

As a result of SDA's efforts to provide sound factual information to EPA, these two major uses of VOC in the cleaning products industry, i.e., hand dishwashing and laundry detergents, did not appear in EPA's proposed regulatory agenda for consumer and commercial products. Twenty-four categories of consumer products are included in the first group of products to be considered for regulation by EPA. However, all 24 categories identified by EPA have already been regulated by one or more states.

This issue is fairly traditional with regard to the scope of ingredients, government interest, and participants involved. The focus is on a large class of compounds having a single attribute—volatility. There is a high level of state and federal activity, but both are driven largely by existing federal law. The participants are regulatory agencies and industry. The VOC issue also illustrates the value of developing and compiling scientific information related to an issue and putting it in a form useful to the other participants, in this case state and federal regulatory agencies.

Antimicrobial/Antibacterial Wash Products

A recent resource-intense issue has been the industry response to the June 1994 Food and Drug Administration (FDA) Tentative Final Monograph (TFM) for over-the-counter (OTC) Health-Care Antiseptic Drug Products. The TFM was issued as part of a regulatory program initiated in the early 1970s. A coalition comprised of SDA and the Cosmetic, Toiletry, and Fragrance Association (CTFA) was formed to respond to the TFM, which would significantly affect antimicrobial general use products.

First, the TFM would eliminate the category of antimicrobial soaps, specifically in consumer products, and include these products under an antiseptic drug category; second, it would establish stringent efficacy testing requirements for all antimicrobial soaps equal to surgical scrubs, including demonstrations of effectiveness against a broad spectrum of microorganisms, fast acting capabilities, and persistence; and third, it would establish new labeling requirements. These requirements would dislocate a $1 billion industry of safe and effective products. FDA's elimination of the antimicrobial soap product category in the 1994 TFM represented a significant departure from its earlier policies, including its own initial treatment of the category in a 1978 TFM.

The industry coalition, recognizing the opportunity for bringing to FDA a reasonable, scientifically sound approach for regulating these products, has developed a Healthcare Continuum Model, new efficacy test methods which have been submitted to ASTM for adoption, and a rationale for accepting the use of a product in a specific category based on the product's intended use. The model demonstrates that specific levels of risk exist in each of the six defined product categories; the size of the population at risk is inversely related to the severity of the risk; and specifically formulated products are the most effective way to manage category-specific risks. Products formulated for particular end uses, whether a surgical scrub, pre-operative skin preparation, antiseptic health care personnel handwash, food handlers' handwash, general use handwash or general use bodywash, assure that appropriate levels and types of active ingredients are being used to address specific risk levels.

One industry concern with the TFM addressed by this model is the potential for inappropriate consumer use of professional strength products to result in negative consequences. The use of professional strength products by the general population could create a near-sterile environment that would allow transient, opportunistic pathogens to replace the normal resident bacterial population on the skin. The Healthcare Continuum Model seeks to ensure that a balance exists between the potential for disease acquisition or transmission and the level and type of active ingredient used.

FDA is considering the Healthcare Continuum Model as a reasonable solution to the proposed TFM. The agency has stated that there is a need for more dialogue between industry, the public, and FDA before the TFM is finalized.

Comments on the TFM were filed with FDA on June 15, 1995. The coalition's efforts have since focused on the data submittal phase of this project, completed December 13, 1995.

This issue is also fairly traditional with regard to the scope of ingredients, government interest, and participants involved. The focus is on a large class of compounds having a single attribute—antimicrobial activity. The scrutiny is at the federal level and is part of a federal regulatory process initiated over 20 years ago. The participants are largely regulatory agencies and industry. Industry response to the TFM also illustrates the value of developing and compiling scientific information related to an issue and putting it in a form useful to the other participants, in this case a federal regulatory agency.

Product Environmental Criteria. The development of environmental criteria to classify cleaning products reflects many of the changes that have occurred in the issues facing the industry. The SDA's overall concern has been that environmental criteria which are not based on sound science will be used to make purchasing decisions or design consumer education programs. These two applications of environmental criteria are described below.

Purchasing Decisions. Efforts have been made to reduce the complexity of safety decisions by establishing environmental criteria to guide cleaning product purchases. Some approaches are designed to affect consumer decisions at the point of sale through the use of a logo. SDA's concerns with these efforts have centered on the lack of sound science in the development of the criteria. Initially, private certification programs were the key players in developing such criteria.

However, the interest of government agencies and, more recently, standard-setting organizations in establishing cleaning product selection criteria based on product environmental attributes presents a more formidable challenge. Participating in these activities to ensure that soap and detergent industry concerns are addressed has been a major SDA effort.

Two government programs aimed at modifying federal procurement of cleaning products by requiring consideration of environmental attributes have the potential to affect our industry significantly. Requirements that products sold to federal agencies meet new environmental purchasing specifications would directly affect formulators selling products in this market.

There is also the potential for an impact on the retail market because the specifications developed in these programs would have the weight of credibility that the public typically reserves for federal programs.

A major issue for SDA with these government programs is the proposed use of existing private certification programs to make judgments on the environmental acceptability of products sold to federal agencies. Such a role would add significant credibility to programs that appear to be arbitrary.

Executive Order. The first federal program stems from President Clinton's 1993 Executive Order entitled "Federal Acquisition, Recycling, and Waste Prevention." Among other things, the Executive Order requires the development and implementation of executive branch policies for the acquisition and use of environmentally preferable products and services.

EPA has the lead in developing guidance for executive branch policies. An EPA draft of the guidance lists the following key principles for identifying environmental preferability:

- Consider environmental preferability early in the acquisition process.
- Consider multiple attributes.
- Address all life-cycle stages, to the extent feasible.
- Consider the scale (global *vs.* local) and temporal (reversibility) aspects of impacts.
- Tailor decisions to local environmental conditions, where appropriate.
- Develop as an important factor in competition.
- Examine product claims closely. Agencies will be directed to Federal Trade Commission guidelines on environmental marketing claims and third-party certification programs.

After issuing this general guidance in final form, EPA intends to develop guidance on specific categories of products, including hard surface cleaners. SDA has engaged EPA staff in a dialogue to discuss both general and specific guidance on cleaners.

General Services Administration. The second federal program started as an initiative of the Public Buildings Service (PBS) of the General Services Administration (GSA) to incorporate environmental criteria in its specifications for cleaning product purchases. PBS manages 7,000 federal buildings. In early 1993, the PBS decided to review its specifications for cleaning products to reduce the number of different products purchased (i.e., reduce 40 down to 5), evaluate product efficacy requirements, and make products that are available to general consumers more accessible to GSA building managers. As part of its specifications review, PBS decided to add criteria for environmental attributes of products to the specifications.

EPA became involved in the program as an advisor, due to the environmental nature of the criteria. The two agencies formed a joint task force to work on the criteria. SDA staff and industry representatives have been interacting with GSA and EPA staff since the fall of 1993.

The GSA/EPA task force completed one study in the fall of 1994. The study evaluated perceptions of maintenance workers and users of a federal office facility in Philadelphia regarding four different groups of cleaning products. One group consisted of products previously used in the building. Three other groups were chosen on the basis of their environmental attributes. The group of products previously used at the facility were perceived to perform best. No meaningful differences in perceived health and safety were found to exist among the four groups. EPA has been evaluating the relative risk of the four groups of products and plans to issue a report of its findings.

GSA's Federal Supply Service (FSS) has now assumed responsibility for the pilot project. FSS has been seeking input from vendors already listed in their multiple supply schedule for biodegradable cleaners/degreasers. FSS is proposing guidance for vendors to supply information *voluntarily* on their products related to a specific list of environmental attributes. Information provided would be included in FSS's description of the products. EPA intends to use information from this pilot project to help develop its specific guidance on hard surface cleaners. SDA has commented extensively on the attributes FSS is considering for the guidance.

President's Council on Sustainable Development. In a third federal activity that could lead to the development of information related to procurement of cleaning products for federal facilities, a demonstration project on product stewardship for building maintenance and cleaning products has been conducted by a project team under the President's Council on Sustainable Development. The objective of the Council is to provide recommendations to the President on how to sustain economic development while improving the environment. Here again, SDA has participated in the undertaking to share its resources and information.

The demonstration project examined how environmental and human safety information on ingredients and formulations is transferred in the chain-of-commerce, using office building cleaning products as an example. The goal of the project was to develop information that could be used as the basis for recommendations on transfer and use of information that could lead to improved eco-efficiency for institutional and commercial cleaning products. The project team's recommendations are to be incorporated into a final report to the President. [The report was published in March, 1996. Ed.]

The project team will be making the following major recommendations:

- Partnerships are required to advance the practice of eco-efficiency of commercial and institutional cleaning products.
- There must be an improved exchange of environmental health and safety (EH&S) information between cleaning personnel and the rest of the chain-of-commerce.
- Building maintenance programs have to be improved to reduce EH&S impacts.
- Regulations which promote eco-efficiency should be nationally consistent, performance oriented, and give appropriate consideration to the life cycle of cleaning processes.

ASTM. Standard-setting organizations have now become involved in this issue. For example, the American Society for Testing and Materials (ASTM) has issued for review and comment a draft Practice for Environmentally Preferable Cleaners/Degreasers. The proposed draft draws heavily on drafts of procurement guidance that GSA has developed and proposals put forward by third-party certification programs. SDA's views on the deficiencies of this effort have been communicated to ASTM.

Consumer Education. Protecting the ability of industry to formulate products that best meet consumer needs and maintaining consumer confidence in the safety of our products are the objectives of SDA activity on consumer education programs. SDA is particularly concerned when the application of environmental criteria to cleaning products in such programs involves classification of products as hazardous products or hazardous wastes. This has been of primary concern to SDA when education programs promote the disposal of cleaning products as household hazardous waste (HHW); or promote mix-at-home recipes to replace formulated cleaning products.

These issues are becoming more problematic because municipal, county, and state governments are developing HHW education programs. For many years, environmental groups have published materials that incorrectly classify household cleaning products as hazardous and promote alternatives to formulated products. This sort of erroneous and unsubstantiated information is being picked up by local governments and included in their educational literature. This development is of concern because consumers regard information disseminated by government as unbiased and more truthful than that of either environmental groups or industry. Further, any inaccurate information distributed by government agencies is difficult to correct and can result in unnecessary cost being incurred by a local government.

SDA believes that accurate information about the safe use, storage, and disposal of household cleaning products is key to successfully addressing this issue. From an educational standpoint, SDA is approaching this in two ways: first, working with local officials to cooperatively develop waste disposal or waste reduction programs, and second, offering existing educational materials at no charge to local officials.

The first approach has already proven to be effective. SDA worked cooperatively with the Wisconsin Department of Natural Resources (DNR) to develop a brochure on waste reduction. The Wisconsin DNR is currently distributing the brochure free of charge to waste officials and educators.

In Los Angeles County, California, SDA is participating in the development of a major HHW public education program. The county has conducted a market research survey to determine what residents know about household hazardous waste. SDA is proposing to local officials and the project's contractor that household cleaning products are neither hazardous nor do they constitute a substantial quantity of household waste; therefore, they should not be included in the HHW education program. Our message is simple, accurate and, we believe, persuasive. Most household cleaning products are:

- water soluble
- designed to be disposed down the drain
- safely treated in sewage treatment facilities
- not a cause of adverse effects in the environment
- only minimally disposed.

If a segment of cleaning products is included in a government education program, SDA is recommending that the focus be on teaching consumers to read and follow label directions, buy only the amount of product they can use, use it up, or give it away. If a product must be disposed, do it properly.

Using SDA's second approach of providing existing materials, we are now offering free of charge to waste officials and educators nationwide our *What Can I Do?* source reduction/disposal brochure. Over 130,000 copies have been distributed since June, 1995.

SDA legislative and regulatory efforts on this issue focus on opposing government-sponsored "information" programs which misrepresent the safety and environmental characteristics of cleaning products and promote so-called "alternative products." At the same time, SDA supports programs promoting accurate information regarding the use and disposal of household cleaning products.

SDA's position is based on its view that the role of government is to provide accurate and objective information to consumers rather than to express or imply a preference for one product, or set of products over another when those products otherwise meet all statutory and regulatory requirements. This is especially true when such programs are scientifically unsound and present misleading information.

These initiatives relative to product environmental criteria illustrate the changes that have occurred in the issues facing the cleaning products industry. These criteria are being applied in pollution prevention programs which do not incur the administrative burdens of regulatory approaches. Instead of focusing on a single ingredient or class of ingredients, these programs tend to lead to scrutiny of many ingredients. These programs have also become popular at many levels of government. In addition to regulatory agencies, these programs involve public interest groups, consumer educators, professional organizations, academicians, and end users.

Although the development of sound technical information will continue to provide a critical foundation for addressing these and other issues that the industry will face, care must be taken to be sure that the information is in a form that is useful for the various participants. While SDA has a high level of credibility among many of these participants, we are expanding our efforts to communicate the extent of SDA's expertise on products and ingredients to all key stakeholders.

The following are additional issues currently involving cleaning products.

FIFRA Reform

For the past several years, industry has seen consumer interest in disinfectant products increasing. However, manufacturers have been hamstrung in their ability to address this growing interest by the fact that under the Federal Insecticide, Fungicide and Rodenticide Act (FIFRA), disinfectant and sanitizing cleaning products are considered "pesticides." As a consequence, these products are subject to the same regulatory processes at both the state and federal level as agricultural pesticides. At the state level, for example, disinfectants were being involved in proposals directed at agricultural pesticides for special fees, container disposal restrictions, and designation as household hazardous waste simply because of their federal designation as pesticides.

At the federal level, these antimicrobial products have become a low priority within the EPA approval process. Despite enacted legislation allowing the EPA administrator to provide for different regulatory treatment of non-agricultural pesticides, which could allow for expedited processing of applications for these products, the EPA had taken no such action. This inaction, combined with reduced resources resulting from budget cuts and a Congressional mandate to re-register all pesticides, has essentially brought the regulatory approval system for antimicrobials, including disinfectant and sanitizing products, to a halt.

In response, SDA joined with the Chemical Manufacturers Association (CMA), the Chemical Specialties Manufacturers Association (CSMA) and the International Sanitary Supply Association (ISSA) to develop and promote legislation to accomplish two broad goals: first, to distinguish antimicrobial pesticides from agricultural pesticides by definition, and second, to assure that antimicrobials receive timely and appropriate regulatory treatment. As a result of this collaboration, legislation that would provide substantial remedy is now being considered in Congress. The legislation contains a definition of "antimicrobial pesticide" which distinguishes these products from other pesticides. Further, it requires EPA to review and streamline its current application and approval processes for antimicrobials. Failure by the EPA to meet the stipulated conditions would result in the imposition of a series of default deadlines. This provision addresses what is perhaps the most critical defect in the current system, the lack of any certainty for the completion of a regulatory action.

A number of other issues are also addressed in the legislation, including the following:

- An exemption for antimicrobial containers from pesticide residue standards.
- A process for the synchronization of federal and state data requirements when making an application.
- Allowance for manufacturers to provide safety information and cautionary warnings for a product diluted for use as well as its concentrated form.

Nonanimal Testing

Due to interest in legislation and regulatory reform initiatives directed at the use of animals in safety testing, SDA continues to actively explore alternative test methods and maintain a dialogue with regulatory agencies. Our research has provided a better appreciation of the applicability of candidate *in vitro* tests to the range of products and ingredients of the detergent industry. This information has been shared with government programs directed toward improving safeguards against eye injury while reducing the use of animals. The data indicate that while a number of alternative tests developed to replace the Draize eye irritation test are useful for screening the eye irritation potential of nonalkaline cleaning products, they vary greatly in their ability to identify the eye irritation potential of alkaline or oxidizing materials.

SDA is building on the data gathered in these validation studies by sponsoring an analysis of the statistical variability of the Draize eye irritancy data. Such variability will be a key element in any future efforts to validate *in vitro* eye irritation assays.

Most recently, SDA completed a review of the scientific literature on all available alternative tests which have used soap and detergent ingredients or products. This is the first such effort by any organization to assemble this type of information. The findings of this review are consistent with the results of SDA's previous studies, showing that there is a wide divergence of opinion as to which assays, if any, are the best predictors of potential eye irritants; that the *in vitro* battery of tests that will most accurately predict ocular irritation will most probably have to be tailored to the class of compounds to be tested; and that there is no single *in vitro* assay capable of replacing the Draize eye irritation test.

Nonanimal tests have proven useful as screening tools, thereby reducing the amount of animal testing. However, because neither the mechanisms of eye irritation nor the mechanisms of *in vitro* tests are completely understood, unconditional validation of any given nonanimal testing approach may not be possible in the near term. Therefore, SDA will be concentrating on particular aspects of the evaluation of eye irritation potential for further investigation and continuing to assist regulatory authorities in their assessment of alternative tests.

Phosphate

Legislation affecting the phosphate content of cleaning products has slowed over the past few years due to the declining usage of phosphate compounds in laundry detergents. Phosphate usage in automatic dishwashing detergents has attracted some attention, but no serious threats have materialized.

As part of an outreach program, SDA has been sponsoring a one-day seminar for the past five years on processes for nutrient removal from municipal wastewaters. These seminars are co-sponsored by local water pollution control associations and state environmental agencies. The hard-cover book *Phosphorus and Nitrogen Removal from Municipal Wastewater: Principles and Practice,* which is the manual for the course developed by SDA, has been in high demand within the engineering community.

Heavy Metals

Because municipal wastewater dischargers continue to exceed permitted limits for heavy metals in spite of full implementation of industrial pretreatment programs, other sources of metal contribution to the collection systems are being scrutinized. Because some historical studies reported in the literature implicated household products as significant sources of heavy metals, a number of municipalities were considering consumer education programs which would have recommended the use of specific brands of household cleaning products based on the reported heavy metal concentrations in the products. SDA realized that the education programs should have updated information based on state-of-the-art analytical methodology applied to current products.

Studies of nine heavy metals in laundry and dishwashing products conducted by SDA have shown that these product categories contribute less than 0.5% of total amounts of any of the metals found in municipal wastewater. This information has convinced government agencies that education programs on these products for the purpose of affecting heavy metal levels in municipal wastewater are not warranted.

FTC Environmental Marketing Guidelines

In 1992, the Federal Trade Commission (FTC) issued "Guides for the Use of Environmental Marketing Claims." This was in response to the urging of SDA, as well as other industry associations, to adopt national guidelines to protect the transfer of truthful information from product manufacturers about the environmental impact of products and packaging while protecting consumers from misleading or unsubstantiated environmental claims. As expected, the guidelines have reduced the proliferation of state and local regulations of environmental claims. Conflicting requirements in different parts of the country would have discouraged many companies from providing any meaningful environmental information to consumers.

In publishing its original guidelines, the FTC indicated that it would review the issue in 1995. FTC is soliciting comments this fall.

Boron

The EPA is considering issuing a proposed drinking water standard for boron as part of a mandate under the Safe Drinking Water Act to regulate 25 new contaminants every three years. In late 1993, the World Health Organization (WHO) issued a drinking water guideline for boron of 0.3 mg B/L, which is considerably below existing limits and guidelines. There has been concern that the WHO guideline is inappropriate and could influence EPA. EPA originally was considering two standards for boron in drinking water, 0.6 or 1.0 mg B/L. However, as a result of the President's Regulatory Reinvention Initiative, EPA has held stakeholder meetings to determine the necessity of regulating boron in drinking water. As a result, the agency is now considering a possible drinking water level of 2.0 mg B/L, as well as reevaluating whether it should regulate boron in drinking water at all.

Because of the potential impact on laundry products of an overly restrictive boron standard, SDA sponsored the preparation of a human health risk assessment on boron which has been completed and submitted for publication. The results of the risk assessment demonstrate that the guideline established by WHO and the limitations being considered by EPA are unnecessarily low. SDA has also compiled and evaluated data on the environmental concentrations of boron in U.S. surface waters. The evaluation demonstrates that boron concentrations in U.S. drinking waters would not be expected to pose any health risk. On the basis of these new evaluations, SDA has concluded that inclusion of boron on EPA's list of 25 substances to be regulated is not justified.

These efforts are being coordinated with groups in Europe, such as the Association Internationale de la Savonnerie et de la Detergence (AIS) and the European Center for Ecotoxicology and Toxicology of Chemicals (ECETOC), which are also addressing boron issues related to drinking water.

Emerging Issues

The following are issues on the horizon. These issues are being addressed by SDA technical and communication programs targeted at the audiences involved. It is hoped that by compiling the necessary information on these issues now and putting it in forms useful to the other stakeholders, these emerging issues will not become the subject of broad public debate.

Enzymes

Issues raised in the early 1970s concerning the use of enzymes in detergents were promptly addressed by the soap and detergent industry to the satisfaction of the concerned parties. As a result, enzymes have been safely handled in manufacturing and in consumer products for years. The recent increased use of enzymes in nondetergent consumer products prompted consideration of the need to reemphasize the use of proper manufacturing techniques and proper practices in occupational settings. SDA has been promoting safe practices in the use of enzyme technology through the recent publication of a manual entitled "Work Practices for Handling Enzymes in the Detergent Industry." This document provides a comprehensive occupational health and safety program which may be used in the detergent industry. It describes enzymes and situations in the manufacturing process in which exposure may occur. It details the potential health effects and the necessity for implementing proper engineering and administrative controls and, if need be, instituting a protective program for personnel. Additionally, it provides a model medical surveillance program detailing measures an employer can take to ensure proper health and safety in the workplace and minimize the potential for cases of sensitization among employees.

The document has been distributed widely in the soap and detergent industry as well as to other interested industries, associations, and ventilation and medical consultants dealing with enzyme handling issues.

Endocrine Disruptors

Certain synthetic and natural compounds are known to have estrogenic properties. Recent scientific studies have associated environmental exposures to these agents with adverse effects that could be caused by their estrogenic properties. Such exposures have been hypothesized to cause endocrine system disruption which may result in reproductive, teratogenic, or carcinogenic effects in humans and animals.

The spectrum of circumstances that may exhibit this type of activity remains ill-defined. EPA has recently reported that through the use of a computer model it has identified 300 chemicals that merit further study for possible hormonal activity. To date, this list is not yet publicly available.

EPA recently conducted a research prioritization and development workshop related to environmental estrogens and endocrine disruptors. Internationally, research workshops have also been held in England, Denmark, and Germany. The recommendations of the EPA workshop were to focus research on plausible, testable hypotheses, and to test these hypotheses using an appropriate degree of rigor. Additionally, it was recommended that chemicals and chemical classes be prioritized for evaluation based on exposures and tonnages used, as well as structure activity relationships (SAR). Finally, it was recommended that EPA develop a tiered testing strategy that utilizes both *in vitro* and *in vivo* assays.

Internationally, the British workshop recommended that chemicals and classes of chemicals be identified for analysis as well as the extent of exposure to these chemicals. The British workshop also recommended the development of a pan-European database for semen quality and quantity. Additionally, the risks to fish and reptiles should be identified. The Denmark workshop recommended the development of rapid, cost-effective test methods for the assessment of estrogenic activity. Finally, the German workshop recommended a reduction in the use of alkylphenol ethoxylates, the initiation of an Organization for Economic Cooperation and Development program to address environmental estrogens, and that concentration limits be established for estrogenic compounds in drinking water.

SDA is monitoring technical and regulatory activity as it relates to environmental estrogens and endocrine disruptors. As with other issues, SDA will be examining ways to support the development of useful information on this issue.

Septic Tank Systems

Twenty-five percent of the U.S. population relies on the use of septic tank-tile field systems for the treatment of their household wastewater. The installation and operation of these systems are loosely regulated. It is believed by some that these systems may be significant sources of contaminants to ground water.

SDA has been gathering information from an experimental field site at a home in Florida to assess the fate of detergent ingredients in soil and groundwater from septic tank systems. Information on the biodegradation rate and travel distance of surfactants in soil and groundwater will be developed using data obtained from a monitoring well network installed at the site. Such information will be required to assist regulatory agencies in making rational decisions about detergent-containing discharges that may potentially affect groundwater quality.

In a related activity, the SDA is sponsoring development of a computer model which can be used to predict the fate of detergent ingredients in septic systems. The model will be useful as a screening tool to assess the fate of cleaning product ingredients in septic tank systems.

Greywater

Water reuse regulations or guidelines have been developed by 36 states, with most regulations or guidelines focusing on the reclamation of wastewater for urban and agricultural irrigation. Increasing demand on municipal water and wastewater treatment systems will promote the growth of water conservation and reuse programs, which can be expected to result in further new legislation and regulation. Additionally, as these programs gain acceptance by the public, existing reuse regulations will be modified, expanded, and tightened. Because greywater reuse could represent a significant source of environmental and human exposure, it is important that the environmental fate and effects of cleaning product ingredients in water reuse systems be characterized.

Greywater is defined as laundry and bath wastewater. Dishwashing wastewater, both from machine and hand operations, is not suitable for greywater reuse due to high grease, oil, and solids content resulting from food wastes. Recently, the California Department of Water Resources issued greywater guidelines that generally regard greywater as unsafe for subsurface irrigation and suggest ingredients that should be avoided. These recommendations are based upon work done at the University of Arizona which analyzed laundry wash products for their alkalinity, conductivity, and boron and sodium hypochlorite concentrations. The Arizona study recommends the purchase of products low in these characteristics.

SDA has responded to this issue by initiating literature reviews to determine the public perceptions and issues associated with greywater and the human and environmental safety impacts of detergent ingredients in greywater reuse systems. The results of these reviews could be used to develop programs to educate consumers and assist regulators as they develop greywater reuse programs and guidelines.

Surfactants in Sediments

Over the last several years, EPA has been in the process of developing sediment quality criteria regulations which will establish limits for concentrations of chemicals allowed in sediments. Similar to water quality criteria, these regulations will be used by EPA to judge when remedial action must be taken.

This development is of interest to the soap and detergent industry because surfactants tend to attach to particles in the environment. These particles settle to bottom sediments of lakes, estuaries and rivers, carrying the surfactants with them. As a result, relatively high concentrations of surfactants can be found in sediments. However, because these surfactants are bound to the sediments, they are not readily available for uptake by organisms living in the sediments. SDA has undertaken a program to determine the biological significance of sediment-sorbed surfactants in an effort to aid EPA when surfactant criteria are considered.

A model has been developed that relates adsorption of anionic surfactants to characteristics of the surfactants and the substrates. Research examining the effect of the organic content of sediments on anionic surfactant bioavailability has been completed. The research demonstrated the mitigating effect that adsorption onto sediments had on surfactant bioavailability.

Following discussions with EPA, SDA is designing a project to survey sediments in the United States to determine which, if any, are of concern due to the presence of surfactants. This information would be developed to assist EPA in its sediment quality criteria program.

Chemical Use Inventory

In 1994, EPA initiated a process to develop a Chemical Use Inventory (CUI). EPA views this initiative as an important data-gathering step to support pollution prevention programs. Such an inventory would require the reporting of chemical end uses. Of particular concern to our industry is the fact that it could subject the chemicals used in consumer products to more public scrutiny without providing corresponding perspective on risks.

EPA is considering implementation of a CUI by modifying the Toxic Substances Control Act (TSCA) Inventory Update Rule (IUR) and/or the Toxics Release Inventory (TRI). Currently, manufacturers of chemicals have to report updated information to EPA every four years under the IUR. EPA is considering amending the regulations to require that chemical manufacturers report end uses, in addition to making a number of broad changes in the IUR. Among the end uses reported would be uses in consumer products. In the future, the agency intends to examine the possibility of modifying the TRI to include facility-specific exposure and use information.

EPA has been exploring the possibility of entering into a regulatory negotiation process with affected and interested parties. Industry representatives participated in an EPA workshop in April, 1995 on proposed modifications of the TSCA IUR, which could be directed toward a CUI. The primary concerns of SDA are the burden of reporting information on a wide range of substances and the release of results to the public by EPA without corresponding information on the safety or risks of the substances.

Conclusion

Sound scientific information provides the foundation for the soap and detergent industry's commitment to human and environmental safety. While current issues affecting our industry present many challenges, they also offer the opportunity to share that commitment with key stakeholders. These stakeholders are more varied than ever. Therefore, increasing care must be taken in the future to be sure that this information is provided to these stakeholders in a timely manner and in a form they will find useful. We at SDA believe that evidence of our industry's commitment has and will continue to successfully address many of the concerns about cleaning products.

Acknowledgments

The author wishes to acknowledge the contributions of other members of the staff of The Soap and Detergent Association in preparation of this paper, including Jenan Al-Atrash, Director, Human Health & Safety; Alvaro DeCarvalho, Assistant Technical Director; Janet Donohue, Director of Communications; Dennis Griesing, Director, Public Affairs; and Jane Meyer, Director, Consumer Education.

Chapter 2

Surfactant Challenges for 2000 and Beyond

David J. Kitko

The Procter & Gamble Company, Miami Valley Laboratories, Cincinnati, Ohio 45253-8707

The presentation of this topic is from a formulator's viewpoint with an emphasis on the consumer product side. It also reflects the views of one who returns to the North American scene after a four-year absence and perhaps senses the changes more acutely. Four subject areas are covered:

1. Recent changes in the surfactant area across the four major products categories—laundry, fabric softeners, dishcare, and hard-surface cleaners;
2. A look at some recent trends in consumer habits and pending changes in washing machine design;
3. A review of the rapid acceleration in surfactant R&D as measured by patent activity;
4. Some views on the evolving role of surfactants in laundry products as other cleaning technologies are brought to the market place.

Recent Trends

Laundry Detergents

In mid-1991, compact granules were completing their sweep of the granular detergent subsegment, linear alkylbenzenesulfonate (LAS) was under fire due to dialkyltetralin impurities which were thought to be nonbiodegradable, methyl ester sulfonates were a hot topic for formulators as the idea of natural and renewable surfactants emerged, and polyphosphate was on the bubble. Since that time, polyphosphate has gone, a second generation of compact granules and compact heavy-duty liquid detergents (HDLs) have appeared, and several new laundry surfactants have emerged. Lever introduced the "structured liquid" executions and even more compact granules. These initiatives have not fared well in the market place, and Lever has returned to more conventional HDL products and lower density compacts.

Today, compact granule formulators continue with three basic surfactant strategies, i.e., LAS/alcohol ether sulfate (AES)/alcohol ethoxylate, LAS/alcohol ethoxylate, and alcohol ethoxylate alone. In granules, surfactant innovations have focused on process changes to improve solubility, physical properties, and performance. The only significant surfactant active change was in the fabric softener subsegment; in that area, P&G has moved away from DTDMAC/fatty alcohol prills to a tertiary amine/organic acid salt for softening and static control in the Bold product.

In the HDL area, the surfactant strategies remain similar to granules, but with greater use of alcohol ether sulfates and alcohol ethoxylates in conjunction with LAS. This dependence is driven largely by the absence of cost-effective builder technology for liquids and formulation restrictions. P&G has introduced a new nonionic surfactant, cocoyl *N*-methyl glucose amide, which has replaced LAS in liquid formulas. This surfactant has a more compact head group than conventional nonionics, promoting stronger interactions with alcohol sulfate and alcohol ether sulfates in contrast with alcohol ethoxylates in which the interaction is weak. It has an intrinsically high Krafft boundary and therefore is best utilized with anionics. The key design criteria for HDL surfactant systems remain hardness tolerance, formulation ease, and strong cleaning on oily soils.

Overall, for surfactants in the laundry area, LAS remains the dominant active. It will likely continue in this role over the rest of the decade because of its cost effectiveness and excellent solubility in cooler water. New materials will continue to emerge and cost effectiveness will determine their success or failure in this segment.

Fabric Softeners

Looking at the fabric softener category, innovation in the surfactant area has centered primarily around reduction and/or removal of difatty alkyl dimethyl ammonium salts and their replacement with more readily biodegradable actives which can be formulated into high active concentrates in liquid fabric softeners. The new materials contain amide and ester linkages which hydrolyze in the environment, leading to more rapid biodegradation. Softening and static control have generally been maintained with some loss in mass effectiveness. Strippability, a key laundry concern, remains similar to past systems. Future innovation in the liquid fabric softener actives will likely continue to focus on the balance of fabric softness, static control, and absorbency. There have been no major innovations in the active systems of dryer sheets over the recent past, and static control benefits and fabric freshness remain the drivers of consumer acceptance.

Hand Dishwashing Liquids

In the hand dishcare category, in which surfactants are the primary actives, the key consumer drivers are grease cleaning, mildness, and suds. Since the introduction of Ivory Clear in the early 1990s, Colgate-Palmolive and P&G have chosen two different approaches to see who can capture the hearts of consumers with new surfactant technology. Colgate has reformulated to use alkyl polyglycosides as a major surfactant active to provide enhanced grease cleaning and mildness with a good suds profile. P&G has recently introduced cocoyl *N*-methyl glucose amide in its reformulated and more concentrated dishwashing liquids. It will be an interesting few years as these new formulations settle into the market place. In the automatic dishwashing product segment, surfactants play the smallest role in all of the cleaning product categories due to the inability to control foam in the machine. Nonetheless, specialty polymeric surfactants are used at low levels as critical components of the formulations to control sudsing and to promote sheeting action to eliminate water spots on drying. R&D efforts here are most often led by suppliers, and new materials are often required as formulations change. This trend will continue.

Hard-Surface Cleaners

The hard-surface cleaner category has perhaps the broadest spectrum of surfactant deployment, ranging from cationics in disinfecting products to specialty nonionics in floor and wall cleaners to bleach-compatible surfactants in bathroom cleaners. The key consumer drivers remain strong cleaning on the key soils—greasy kitchen dirt, soap scum and lime scale, and bacterial kill in bathrooms, all with great shine as the desired end result without excessive rinsing. The new surfactant technology emerging is the use of short-chain surfactants, both nonionic and anionic, to provide full strength and dilute cleaning with great shine as end results. These materials are replacing solvents as environmental concerns rise and regulations on volatile organic compounds tighten.

Trends in Consumer Habits

Let us consider briefly the changes that are occurring in consumers' habits as well as the pending changes in washing machines and the challenge they bring to surfactant technology. Research conducted by P&G in 1993 compared with similar research conducted in 1988 suggests the following:

1. Soil levels are similar, and loads are mostly moderately or lightly soiled (~90%).
2. Load sizes remain unchanged.
3. Water levels are similar.
4. Wash temperature has dropped, with the mean at about 88°F.
5. Chlorine bleach usage is down slightly.
6. Fabric softener usage is similar over the period.
7. Detergent is frequently underdosed.
8. Wash frequency is similar.
9. Dryer drying predominates.

In the area of washing machines, current vintage machines continue to offer increased capacity, but this has not made a significant change in consumer habits. The pending machine design changes, however, do represent several challenges for formulators and these have been well articulated in recent articles, e.g., INFORM, January 1995, and are the subject of a separate section in this conference. My opinion is that surfactants will have to dissolve and clean well in colder water, foam less, and if the water to fabric ratio falls, they will have to suspend removed soils better. Real technology optimization for detergents will occur as the new machines enter the market and a new set of consumer needs emerges. Regardless of machine design, however, improved cleaning in cooler water has been a steady trend that will allow opportunities for surfactant innovation.

Patent Trends

It is also useful to analyze patent trends in the laundry area. These data come from an internal review of patent applications filed globally over the '92–'94 time frame. Patent activity over this three-year period has been intense, with over 1,500 applications filed (see Table 2.1). The dominant sectors are surfactants with over 33% of the applications, bleach technologies with approximately 25%, and enzymes at about 15%. Process patents for granules, polymers, and builders are about equal at 9% each. It is clear that a substantial amount of R&D activity in the laundry area is focused on surfactants. A more in-depth analysis of patents in the surfactant area indicates that there are spikes in the area of naturally derived surfactants, formulations employing higher levels of nonionic surfactants, and development of high active surfactant particles with the ultimate represented by Shell's dry powder form of secondary alkyl sulfates. Because commercialization always lags behind invention, this activity level strongly suggests that surfactant innovations beyond what was covered earlier in this chapter, will occur in the second half of the decade.

Future Surfactant Requirements

What does the recent past suggest will happen in the future for surfactants?

1. Significant improvements in meeting consumer needs through surfactants will lead to commercialization of new materials, e.g., alkylpolyglycosides and cocoyl *N*-methyl glucose amides;
2. Environmental safety improvements and legislation can be a driver in changing surfactant technology, e.g., changes in fabric softener actives and evolution of short-chain surfactants in hard-surface cleaners;

TABLE 2.1 Intense Patent Activity in Cleaning Products—Analysis of 1500 Applications Filed 1992–1994

Sector	%
Surfactant area	33
Bleach technologies	25
Enzymes	15
Granules process	9
Polymers	9
Builders	9

3. Cost effectiveness will continue to be the deciding factor in new surfactants;
4. Peripheral benefits such as naturalness and renewability without cost effectiveness will remain a dead end, e.g., methyl ester sulfonates.

Evolving Role of Surfactants in Laundry Detergents

It is clear that surfactants will not only remain the primary cleaning ingredient in laundry detergents, but may increase in importance with the passing of polyphosphate. Surfactants or surfactant combinations must pick up some of the role of this multifunctional ingredient. Current builder technologies simply sequester hardness. Peptization and soil suspension or antiredeposition will become critical attributes for surfactant systems. This is particularly true as changes in the North American wash process and globalization of laundry products require technologies to work under higher soil loads. Compatibility or synergy with other key cleaning technologies will also become a key design criterion as well as a point of invention as detergent formulators work to use enzymes, bleaches, and polymers to enhance cleaning and provide additional benefits in their products.

By way of prediction and speculation—one or more new surfactant technologies will be deployed in North American laundry products in the last half of the decade. LAS will continue as the workhorse surfactant for granules, but suppliers and formulators will continue to work to overcome its deficiencies with cost-effective alternatives.

Chapter 3

A New and Promising Chapter in the Sodium Silicate Story

Hans-Peter Rieck

Hoechst AG, 65926 Frankfurt/M., Germany

Conventional Silicates

Silicates have traditionally been employed in detergent formulations (1). There are the *soluble* amorphous silicates, or water-glass species, and the *soluble* crystalline metasilicate. The amorphous silicates have been used for many decades in home laundry formulations, first in combination with soda ash as a precipitating builder, later as a source of alkalinity and as a buffering agent and corrosion inhibitor. On the other hand, the highly alkaline metasilicate is widely used in automatic dishwashing and I&I laundry detergents.

The *insoluble* sodium aluminosilicates, or zeolites, have been widely employed as detergent builders in several regions of the world and have emerged as the preferred tripolyphosphate substitutes. The *insoluble* clay type silicates bentonite and hectorite tend to be used, alone or in combination, with cationic surfactants or tertiary amines, for fabric softening. In addition, there is a new group of layered sodium silicates that *dissolve slowly* in aqueous systems and have properties in between the above-mentioned substances.

Zeolite A became popular because it could contribute to solving the eutrophication problems of phosphate-containing detergents. In many parts of Europe, as well as in Japan and in the United States, sodium tripolyphosphate (STPP) in household detergents has been replaced in the last ten years by a system consisting of zeolite, soda ash, and polycarboxylates (Fig. 3.1).

However, the worldwide successful advance of zeolite A has slowed down. While the replacement of STPP by zeolite is virtually complete in Japan and in the United States, in Western Europe, after a significant increase in zeolite A consumption for detergent application from 1986 to 1992, demand has been stagnant since 1993 at 520,000 metric tons per year (Fig. 3.2) (2). This has resulted in around 300,000 tons of surplus capacity and has pushed prices down.

Unexpectedly, the decrease in Western European STPP consumption leveled off at around 300,000 tons per year (Fig. 3.3) (2). The reasons include a modified environmental assessment of STPP and zeolite which has diminished their differences and a slower increase in conversion to compact type detergent powders.

Nevertheless, some companies continued research and development work to find alternatives to STPP and zeolite A, aiming at builder systems that meet the require-

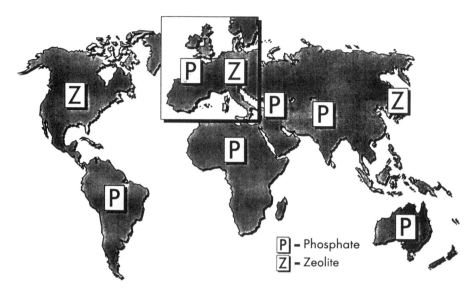

Fig. 3.1. Use of builders 1995.

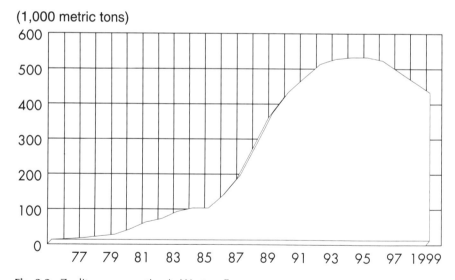

Fig. 3.2. Zeolites consumption in Western Europe.

ments of today's and tomorrow's powder detergents even better. One of the new builders is the aluminosilicate zeolite MAP, which was introduced to the European market in 1994 in a superconcentrated heavy-duty laundry detergent.

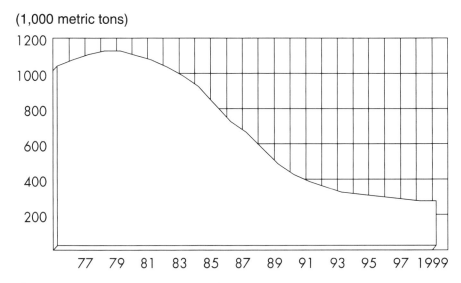

Fig. 3.3. STPP Consumption in Western Europe.

Recently, however, there have been voices in some countries favoring a return to soluble builder systems to minimize the build-up of deposits in sewers, rivers, and lakes and to prevent any increase in sludge quantities at sewage treatment plants. Various sodium silicate systems are offered for this purpose: cogranules of sodium silicate with soda ash, amorphous hydrous sodium silicates, and crystalline layered sodium silicates. Some European producers of these materials are obviously quite convinced of the prospects for their products.

Cogranules of sodium carbonate and sodium silicate are offered to the market as an easy-to-use material. The manufacturer describes the following features:

- Complete and rapid elimination of calcium and magnesium;
- High dispersing power;
- Better safeguarding of the environment because of no sludge in water treatment stations.

The cogranule has a composition of 55% soda ash, 29% (amorphous) disilicate and 16% water. It can be viewed as a silicate solution absorbed in soda ash.

Hydrous amorphous sodium silicates have been promoted for many years as co-builders to tripolyphosphate and zeolite. Those spray-dried solid sodium silicates can be post-added in granular form and are especially suitable for non-tower processes. They dissolve quickly in water and react similarly to water-glass solutions. Some of the features as described by one of the makers include:

- High bulk density;
- High detergency;

- Fast solubility;
- Corrosion prevention;
- Excellent buffer capacity.

Under hard and very hard water conditions, hydrous sodium silicate binds calcium water hardness by calcium silicate precipitation. The reaction proceeds well in the presence of high concentrations of calcium and silicate ions, whereas under soft and very soft water conditions the effect is poor (Fig. 3.4).

Silicate producers aim at improving the calcium sequestration by increasing the degree of polymerization and molecular weight distribution of the silicate species. Nevertheless, most amorphous sodium silicates have been used only because of their reserve alkalinity and anticorrosion properties, and other builders such as STPP or zeolite take over the function of water softening. Sodium silicates can be characterized by the ternary diagram of Na_2O, SiO_2, and H_2O (Fig. 3.5). The upper corner represents 100% water, the right corner silicon dioxide, and the left corner sodium monoxide. Sodium hydroxide, anhydrous sodium silicates and silicic acids occupy positions along these three axes. The region of commercial solutions ranging from 50 to 60% water is indicated on the diagram. Also indicated is a group of sodium silicates with a layered structure, which includes Kanemite and Makatite. These silicates are polymeric crystalline substances with interesting properties for detergent application. The d-disilicate is anhydrous—hence its position is on the bottom axis. This species is being used in several European and Japanese home laundry powder detergents.

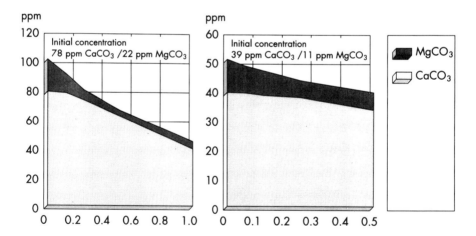

Fig. 3.4. Residual water hardness after addition of amorphous disilicate [pH 10.0/25°C/10 min].

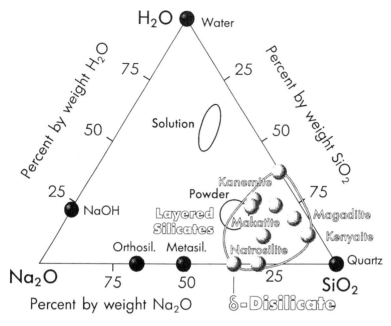

Fig. 3.5. Ternary diagram Na_2O-SiO_2-H_2O.

Layered Silicate

In contrast to the well-known water-glass or amorphous disilicates with a molecular structure consisting of various disordered relatively small silicate rings or chains, δ-disilicate has a polymeric inorganic structure with a very regular arrangement (Fig. 3.6) (3). Due to its polymeric character, δ-disilicate dissolves slowly only in aqueous systems, especially when water hardness ions are present.

When washing with normal tap water, the sodium ions are rapidly replaced by the calcium and magnesium ions which cause the water hardness of commercial δ-disilicate. This occurs before any appreciable δ-disilicate dissolution takes CAN TAKE place. Consequently, during the wash and at a pH of around 10–11, δ-disilicate remains mainly insoluble in the form of ion-exchanged calcium/magnesium/sodium silicate. During rinsing and in the waste water, the wash solution is diluted and the pH finally reaches a neutral level. Under these conditions, the calcium/magnesium/sodium silicate is moderately protonated and is no longer stable. It releases calcium and magnesium ions, and eventually the silicate dissolves completely in the waste water.

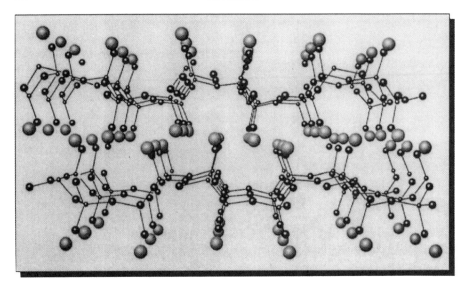

Fig. 3.6. Structure of δ-disilicate.

Table 3.1 demonstrates the various physical phases of builders: the finished powder detergents in the wash solution and finally in the waste water.

Detergent Trends

Besides solubility considerations, the "less is more" aspect is becoming increasingly important in the continued development of builder systems. Multifunctional substances, such as δ-disilicate, which not only bind the water hardness but also provide the alkalinity necessary for the washing process and buffer the pH level as shown in Fig. 3.7, contribute to a further reduction in the required detergent dosage per wash.

TABLE 3.1 Reaction of Builders with Natural Water Hardness and State (solid/dissolved) of the Builder at Various Stages

Builder	Builder-hardness ion complex	
in the finished detergent	in wash liquor	in waste water
STPP	Chelation	Depolymerization by hydrolysis
Citrate	Chelation	Equilibration
Carbonate	Precipitation	Dissolution
Amorphous disilicate	Precipitation	Dissolution
Zeolite A/P	Ion-exchange	Ion-exchange
δ-Disilicate	Ion-exchange	Depolymerization by hydrolysis

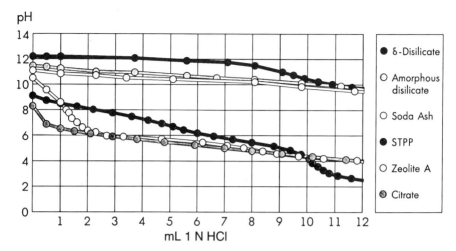

Fig. 3.7. Reserve alkalinity of various builders. Titration of 2 g of builder in 1 L of water with 1 N HCl at 22°C.

Most modern compact-type detergents not only possess a higher bulk density but also contain a higher concentration of active components. Fillers such as sodium sulfate have been nearly eliminated, and the water content has been reduced, resulting in less detergent consumption per wash cycle (Fig. 3.8).

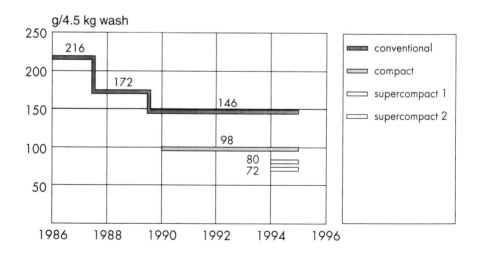

Fig. 3.8. Dosage of European powder detergents (normal soil/hard water).

In 1994, a new generation of superconcentrated detergents with highly active and multifunctional ingredients was launched in Europe. In Japan, the reduction in dosage also continues. New superconcentrated detergents which require just 20 g per washload instead of 25 g were launched in March '95.

Patent Activity

What developments are ahead? First indications of new products can be gained from an examination of patent applications. In February 1995, we analyzed those entries in Chemical Abstracts which contained the terms "builder" and "detergent" in the abstract title. Seventy-six entries could be attributed to powder detergents with a priority date of 1989 or later. Usually, those applications are published one and a half years after they are filed. Nevertheless, those patents or patent applications are often the first indication of innovations.

We tried to assign each entry to a class of builder materials. Occasionally, when several builders were part of the invention, the entry was divided (Table 3.2).

Setting the yearly patents as 100%, we can see that the organic builder patents have a 60% share. They were differentiated into monomeric and polymeric carboxylates. Though still an important focus of R&D activities, especially for biodegradable polymers, organic builders lost their "innovation share," whereas inorganic materials received increasing attention (Fig. 3.9).

Phosphates were not the subject of any of the patent applications, indicating that there is little R&D activity. The producers prefer to concentrate on their production cost and market share rather than spend large sums on research and development.

The aluminosilicate zeolite A was usually named only in combination with other builders such as layered silicates. On the other hand, there were several applications of zeolite MAP clearly showing a new builder activity. Several patents claimed builder

TABLE 3.2 Number of Patents Stating "Builder" and "Detergent" in Chemical Abstracts Title

Builder	1989	1990	1991	1992[a]	1993[b]	Sum
Phosphates	0.0	0.0	0.0	0.0	0.0	0.0
Amorphous silicates	0.3	1.6	0.9	0.8	0.0	3.7
Layered silicates	0.5	1.7	2.5	4.2	1.5	10.4
Aluminosilicates	3.8	3.7	2.7	2.8	0.5	13.6
Carbonate	0.3	0.5	1.4	0.5	0.0	2.7
Monomeric carboxylates	5.5	3.8	3.2	2.0	0.0	14.5
Polymeric carboxylates	10.5	8.7	5.0	6.0	1.0	31.1
Sum	21.0	20.0	15.6	16.4	3.0	76.0

[a]Entries still possible, though not very likely.
[b]Incomplete year.

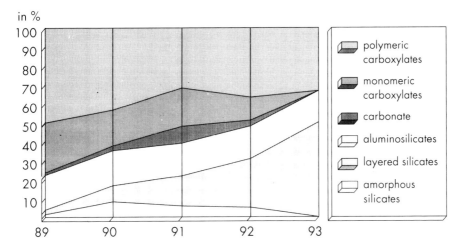

* "builder" and "detergent" in Chemical Abstracts title; application/priority data; 1993 not complete

Fig. 3.9. Builder patents distribution 1989–1993 (priority data).

combinations or detergent formulations with layered silicates focussing on δ-disilicate or similar substances. As Fig. 3.9 shows there was an increasing "innovation share" held by layered silicates.

Carbonates and amorphous silicates also gained attention. In some cases it was not clear whether amorphous or crystalline silicate was the real target of the claims.

Builder Attributes

As already indicated, builders fulfill the functional requirements in different ways: some are very good in binding calcium water hardness, others have high reserve alkalinity or anticorrosion properties or high dispersing power. Some are all-around performers, having multifunctional properties. However, in addition to good performance, their cost should be low and they should be safe and benign to the environment.

Ternary diagrams have been used in an attempt to illustrate the interrelations among cost, performance and ecology (4). The ECP ternary diagram describes the assessment of a material with respect to ecology, cost, and performance (Fig. 3.10). An imaginary substance positioned in the upper corner is ecologically very friendly. However, it is very expensive and has no performance benefits. A substance positioned in the left corner is very cheap, but is detrimental to the environment and has no performance benefits. Finally, an imaginary compound positioned in the right corner shows excellent performance but is very expensive and very harmful to the environment. The real cases are somewhere in between. A material with a position in the center of the diagram is acceptable on all three counts, whereas one positioned near an edge is deficient on at least one count.

Fig. 3.10. The ECP ternary diagram: Ecology vs. cost and performance.

Figure 3.11 shows an attempt to position current builder materials. Sodium sulfate, not normally considered a builder, is very cheap but has virtually no builder effect, although it provides some benefit in the production of crisp free-flowing powders. However, in some countries, the contribution of sodium sulfate to the salt level of rivers is a concern. Water, either bound or absorbed, is environmentally friendly and very cheap, but it has no builder performance. It is usually unwanted, because it decreases the storage stability of bleaching agents, especially percarbonate, and it increases the weight and volume of the finished detergent formulation. However, the quantity of water, bound or absorbed by certain materials, can be reduced only with additional effort and cost. Nevertheless, it is sometimes worthwhile to minimize the water content to improve the stability of hydrolytically sensitive substances.

Soda ash is a traditional and cheap builder material. It provides alkalinity and contributes to the detergency performance, but requires additional builders and antiredeposition agents. Amorphous sodium silicate is more expensive than soda ash. It also provides alkalinity, has good buffering capacity and anticorrosion properties, and can bind magnesium hardness. Like soda ash, it shows good environmental compatibility.

Zeolite A has become fairly inexpensive in several regions. Both zeolites have good calcium-binding capability, improve the flowability of the detergent powder, and are usually ecologically acceptable. However, in some regions, there are some who prefer soluble builders to prevent any sludge increase.

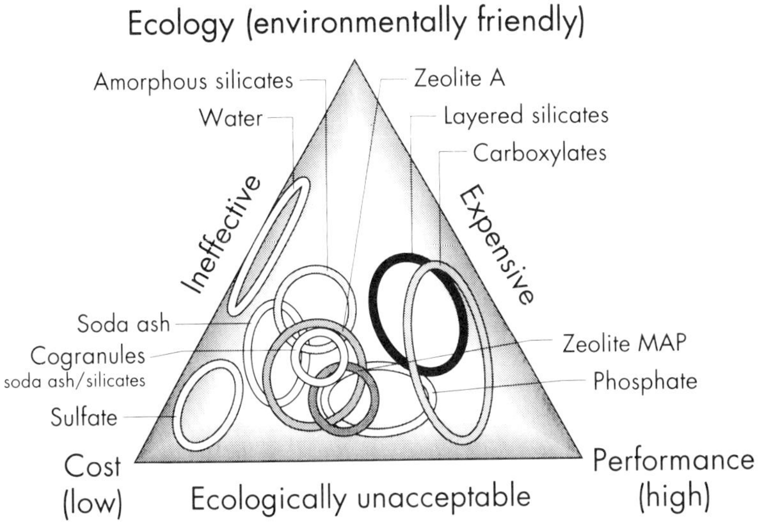

Ecology (environmentally friendly)

- Amorphous silicates
- Water
- Zeolite A
- Layered silicates
- Carboxylates
- Soda ash
- Cogranules soda ash/silicates
- Sulfate
- Zeolite MAP
- Phosphate

Cost (low) — Ecologically unacceptable — Performance (high)

Ineffective — Expensive

Fig. 3.11. ECP triangle of builder, 1995.

Tripolyphosphate has very good builder properties and is available at a reasonable price. In some countries there are no objections to STPP use, whereas in others, there are bans, limits, or STPP is omitted for marketing reasons. In the United States, phosphate has virtually disappeared from consumer laundry detergent formulations, the result of voluntary removal by the major detergent manufacturers. Layered silicates are new builders offering the benefits of multifunctionality. Their price is still comparatively high.

The carboxylate group includes monomeric compounds with several carboxylic groups such as sodium citrate, as well as polymeric materials. Carboxylates are more expensive than the materials discussed previously. Citrate and other monomeric polycarboxylates often exhibit good builder properties, sometimes limited to low and medium temperatures, and have little or no adverse effect on the environment. Polymeric polycarboxylates enhance the washing power of formulations as a result of their effectiveness in dispersion and inhibition of crystal growth. The traditional polymeric polycarboxylates are only slightly biodegradable, but their concentration in the environment is reduced by more than 90% by the sludge in sewage treatment plants. New biodegradable polycarboxylates such as polyaspartate have an intermediate environmental position within the carboxylate range, but are expensive due to relatively high production costs.

The positions of builders in the ECP triangle do not remain static, but are in flux (Fig. 3.12). With increasing environmental awareness in many countries, the ecological aspects are now considered more critical. When we compare the positions of materials some years ago and today, the axis has shifted and we see a less ecological position although counterarguments have been advanced for some materials. Further, the price of compounds is certainly not static, but depends on technological progress and the market situation. Therefore, even if opinions coincide at the present time, it does not mean that this will be the same in a few years.

In principle, only those materials that occupy a position near the center of the diagram will be used on a long-term basis, because they are acceptable in terms of ecology, cost, and performance. It will therefore be the goal of producers to move their products toward that position (Fig. 3.13). For soda ash and sodium silicate, this can be accomplished by producing cogranules that offer benefits in handling and some improvement in detergency performance. In the case of water glass, spray-dried and granular forms are supplied that offer advantages in the detergent manufacturing process. Producers of other materials might aim at improving or protecting the ecological image of their builders, and others will strive to reduce their production costs by installing capacities of suitable size.

Builder acceptance is not just a matter of the property profile. Builders have to fit into the detergent trends with some being more suitable for spray tower routes and conventional detergents and others for non-tower processes and superconcentrated products. Therefore, detergent trends strongly affect builder consumption.

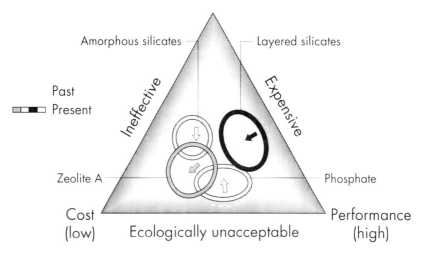

Fig. 3.12. Past and present.

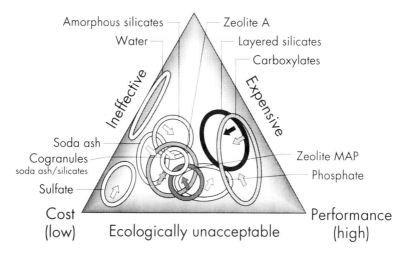

Fig. 3.13. Targets and strategy.

Forecast

We anticipate the following global trends for detergents:

- Further reduction in detergent dosage per wash by use of relatively high concentrations of efficient and multifunctional ingredients;
- Increase in market share of detergents with oxygen bleach and/or highly efficient enzyme systems;
- Solubility of the detergent raw materials in waste water as a long-term target;
- Increase of percarbonate bleach at the expense of perborate;
- New detergent production plants making use of non-tower routes for the manufacture of detergents;
- Continuation of or return to traditional low-cost formulations during economic recessions;
- Advance with efficient and environmentally friendly formulations during periods of economic expansion.

Taking the individual builder properties into account, including cost, performance and environmental friendliness as well as the anticipated detergent trends, we expect the following for builders:

- Decrease in further phosphate substitution by zeolites;
- New aluminosilicates in some premium brand detergents at the expense of zeolite A;

- More multifunctional builders;
- Increase in sodium silicate builders, especially δ-disilicate;
- Some increase of monomeric carboxylates as co-builders;
- More and more biodegradable organic polymers as dispersants at the expense of the current acrylate polymers and acrylate/maleate copolymers;

Recent pressures in favor of environmentally friendly, cost-effective detergents have created innovations in several aspects of the industry. Those developments have not yet come to an end. However, this progress must be affordable. The recent recession of the early 1990s directed attention to cost-efficient formulations, whereas the readiness to move to even more environmentally friendly products is higher during periods of economic expansion. Apart from cost efficiency and environmental friendliness, the market penetration of new builders will depend on the patent situation and the licensing policy of patent holders, both of which affect the availability of the materials.

Builders are an essential and weighty part of detergent formulation. Therefore, even if there is a driving force to change the builder system, it cannot be done overnight. It requires years to install new production capacities on the builder producers' side and to adjust the production process on the detergent manufacturers' side before a system of several brands is completely changed. For this reason, we expect the materials in current use to be the main builders at the turn of the century as well. Nevertheless, technological progress will not come to a standstill; the turbulent times of the present will last for a few more years until, maybe, one or a few builder system(s) emerge to fulfill nearly all of the requirements and satisfy the needs of the detergent industry and, of course, the consumers.

References

1. Rieck, H.-P., Builders: Backbone of Powder Detergents, in *Powder Detergents,* Marcel Dekker, Inc., New York, NY, in press.
2. Smart, M., and R. Willhalm, in *Chemical Economics Handbook, CEH Marketing Report on Zeolites,* SRI International, Menlo Park, 1995.
3. Hoechst High Chem: SKS-6™—Detergent Builder of the Future, May 1993.
4. Rieck, H.-P., in *Proceedings of the 3rd World Conference on Detergents: Global Perspectives,* edited by A. Cahn, Champaign, IL, 1994, pp. 161–167.

Chapter 4

Advances in Detergent Processing

Michael Hill[a] and Robert Ahart[b]

[a]Unilever Research U.S., Edgewater, New Jersey, and [b]Lever Brothers Co., Edgewater, New Jersey

Introduction

The traditional view of the detergent business places all product developments in the hands of the formulators. Indeed, much of the progress in new products has been the result of new materials that have greatly enhanced their performance. For example, fabric washing products have witnessed the emergence of perborate and silicate in the 1920s, synthetic detergents, phosphates, and fluorescers in the 1930s and 1940s, proteolytic enzymes in the 1960s, and bleach activators and zeolites in the 1970s. Nevertheless, it has been said, "You can't sell it if you can't make it!" This underscores the importance of processing in the detergent industry.

The limitations of manufacturing processes have often constrained product formulators. For example, the detergent industry usually accepts only new materials that can be processed in existing equipment. On the other hand, process innovation has provided an additional avenue for new products and product forms. Current trends in detergent processing illustrate these points. The first of these trends has been the emergence of processes that yield detergent powders with high bulk densities. This evolution in the way detergent powders are made has significantly changed the product form. Furthermore, these new processes promise to be a basis of selection for future new materials. A second trend has been increased complexity in the processing of liquid products, a consequence of the move toward products with more complicated physical structures. Today's gels and structured liquid detergents require sophisticated processes. But even more significantly, advances in processing have also resulted in a new product form, concentrated fabric softeners. A third trend is the use of novel processes to protect sensitive ingredients. While this area has seen more patent activity than commercialized processes, it has already led to new products and offers to open new vistas for formulation flexibility.

Processes for High Bulk Density Powders

The first fabric-washing products in a non-bar form were soap powders, appearing as far back as 1870. Around 1910, molten neat soap was poured over a rotating roll to make ribbons, and then dried to form soap flakes. By 1920, inorganic builders were being mixed into the molten kettle soap in the soap crutcher, but the process was unchanged. However, a major advancement in processing came with the introduction of spray driers in the mid 1920s. These converted crutched soap into uniform granules with a low bulk density and high specific surface area, resulting in high rates of dissolution and sudsing (1).

The arrival of synthetic anionic detergents in the 1930s and 40s perpetuated the process of mixing in a crutcher followed by spray drying. The spray-drying process was even deemed essential at the time because of the very nature of the materials. Anionic surfactants were prepared by sulfonation/sulfation followed by neutralization with aqueous sodium hydroxide, thus yielding a paste with appreciable water content. These pastes are difficult to incorporate into detergent powders by dry mixing. However, slurry mixing and spray drying in the conventional soap granule processing equipment produced a free-flowing powder with a low bulk density and high rate of dissolution.

Spray drying continues to be used in the U.S. Nevertheless, spray drying is an energy-intensive process requiring significant capital investment. In addition, volatile components in surfactants often lead to hydrocarbon emissions in the atmosphere, and may require the installation of pollution abatement equipment, further exacerbating the capital requirements. Furthermore, the desire to reduce production and distribution costs has motivated manufacturers to raise the bulk density of detergent powders. Because there is an upper limit in the bulk density that can be achieved through spray driers, the desire for denser powders has prompted the development of new processes for manufacture of detergent powders.

One alternative to spray drying that has long existed is simple dry blending. This process consists of low-intensity dry mixing of inorganic salts with low-viscosity surfactants, which are absorbed into the pores of the salts. Processes of this sort have traditionally limited surfactant choice to nonionic surfactants, because anionic pastes cannot be easily incorporated via this process. This, in turn, has greatly limited the performance of detergent powders made in this way. As a result this process has not been an option for premium performance powders.

A major advancement has been the development of processes that utilize post-tower densification of spray-dried powder. Equipment used in these processes includes the intensive (high shear) mixers manufactured by companies such as Fukae, Lodige, and Schugi. In Japan in the 1980s, this process led to the introduction of concentrated powders, a new product form that has since spread around much of the world. This intensive mixing equipment can also be used to produce high bulk density adjuncts via agglomeration. These adjuncts can be post-dosed into spray-dried powders, giving manufacturers the capability of producing a wider range of product bulk density. They are prepared by intensively mixing components containing high water content with hydratable salts that can absorb water, such as sodium carbonate, sodium sulfate, magnesium sulfate, and aluminosilicates (2). This equipment can even be used to manufacture a complete detergent powder without spray drying. Unlike dry blending, these processes also offer high formulation flexibility because they allow for the incorporation of anionic detergent (3). These non-tower processes are emerging as the preferred route for the manufacture of premium powders in much of the world.

Recent developments in surfactant form have expanded the formulation range for simple dry blending. As mentioned previously, anionic surfactants in their traditional paste form are not easily dispersed or absorbed onto powders in a dry blender, and contain too much water as well. The emergence of solid anionic surfactants allows the manufacture of mixed surfactant powders through dry blending. Similarly, low-moisture liquid surfactant blends of anionics neutralized in nonionic surfactant may be sprayed onto inorganic granules, again producing mixed active powders with premium performance (4).

Processes for Complex Liquid Products

Historically, liquid detergent products were manufactured by simple batch blending processes. Not surprisingly, *Kirk-Othmer Encyclopedia of Chemical Technology* devotes three pages to the processing of detergent powders, but only two sentences to the processing of detergent liquids (5)! Formulations were typically isotropic, Newtonian liquids that were insensitive to process shear. Scale-up generally addressed issues such as foaming and dissolution of solid ingredients, while industry process initiatives focused on manufacturing efficiency.

Given the large scale of fabric detergent manufacture, continuous processing offers many advantages over batch processing, including lower capital and space requirements. While continuous processing is not new in the chemical process industry, in the past many detergent manufacturers had been reluctant to adopt continuous processing. This has been due in part to a lack of confidence in the accuracy and reliability of continuous metering equipment. Significant advances have been made in recent years by the manufacturers of pumps and flowmeters. Accordingly, the detergent industry is converting many liquid processes from batch to continuous operation.

Recent changes in liquid detergent formulations are also making a significant impact on processing. One example has been in the processing of gel and structured liquid detergent products. These are not simple isotropic solutions, but instead have a complex phase structure typically stabilized by polymers. These complex products are often shear sensitive, requiring specialized agitation. For example, the polymer structurant for gel products typically requires high shear during its dispersion, but lower shear during subsequent addition of ingredients. The right order of addition is also crucial for obtaining the right viscosity/shear rate relationship. Similarly, heavy-duty liquids structured with deflocculating polymers require high shear to obtain the desired lamellar droplet size (6).

Advances in processing of detergent liquids have even resulted in a new product form, the triple concentrated Ultra-type fabric softeners. Traditional dilute fabric softeners are a low phase-volume dispersion, and can be manufactured in conventional liquid mixing equipment. However, this equipment is incapable of handling the extremely viscous intermediate gel phase of the triple-concentrated softener formulations.

Triple-concentrated softeners are made by processes with multiple high-shear dynamic mixers and a brine injection system that achieve the target droplet size distribution and the desired product viscosity. These processes combine size effects of the active droplets with the ionic strength of the continuous phase to achieve phase stability (7). Note that it is the process, not some added ingredient, that makes this product form possible.

Processes for Incorporating Incompatible Ingredients

A current trend in detergent formulations has been the incorporation of small quantities of weight-effective, space-efficient specialty ingredients that provide specific performance benefits. Examples of such ingredients include bleaches, bleach activators, specialty polymers, enzymes, and perfumes. These ingredients are often difficult to formulate due to chemical reactivity. This causes profound problems when formulating products that must remain stable for many months. This is frequently encountered when formulating aqueous products. For example, machine dishwashing gels that contain sodium hypochlorite do not currently contain nonionic surfactants due to the incompatibility of the two materials. Formulators of machine dishwashing powders avoid this problem by incorporating bleach as a solid precursor, granular dichloroisocyanurate. This material is less reactive in the dry, solid state but dissolves in water to release the more aggressive hypochlorite. As a result, machine dishwashing powders can contain both bleach and nonionic surfactant, and consequently outperform their liquid counterparts.

One alternative to formulating around reactive ingredients is to isolate them. Dual-compartment packaging affords such an opportunity. In the arena of personal products, the recent success of Chesebrough-Pond's Mentadent toothpaste illustrates how packaging can isolate incompatible ingredients, specifically hydrogen peroxide and baking soda. However, the cost structure of detergent products generally precludes the use of such expensive packaging. On the other hand, encapsulation provides a means of isolating ingredients by providing a barrier that protects the encapsulated material during storage. Not unexpectedly, there has been significant patent activity in this area over the past few years.

The key hurdle to overcome in using encapsulation involves development of a reliable release mechanism. Recent published inventions utilize a wide range of release mechanisms including melt, pressure, solubility, and dissociation. Capsules may be prepared via coacervation, spray, or coating processes. Encapsulation processes are often difficult to scale up and can be expensive. This is part of the reason why encapsulation technology has not been exploited within the detergent industry to the same extent as in the pharmaceutical industry.

One detergent product that exploits encapsulation technology is Lever's Sun MicroGel, currently on sale in France. This aqueous machine dishwashing gel contains suspended wax-encapsulated chlorine prepared via a proprietary spray-coating process (8). Because the chlorine is isolated within a wax shell, the product can contain bleach-sensitive materials such as nonionic surfactant. Accordingly, this product has superior performance to competitor machine dishwashing gel products. Procter & Gamble has also developed encapsulation technology that provides protection and delayed release of perfume in dryer sheets (9). Undoubtedly, we will see further advancements in encapsulation processes in the years ahead.

Future Challenges

Despite the advances in processing in the categories described above, the process to manufacture personal washing bars has changed little over the years. While formulations are moving away from soap toward both synthetic bars and soap/syndet blends, the fundamental process still consists of milling and plodding (extrusion). Yet new materials that are being added to these personal washing bars change the phase structure and make them more difficult to process using standard equipment. These processing problems are currently being addressed by further changes in formulation, such as by the addition of co-surfactants, waxes, or starches as "processing aids" (10). However, this continues to restrict formulation flexibility substantially.

As these examples illustrate, processing continues to play a critical role in the detergent industry. Processing is an essential part of manufacturing products containing new materials and often becomes the limiting technology. In addition, processing comes to the fore when it becomes the means to create new products and product forms. The need for these benefits will present challenges to our process engineers for many years to come.

Acknowledgments

The authors thank Craig Cicciari and David Lang for their contributions.

References

1. Lief, A., *"It Floats"—The Story of Procter & Gamble,* Rinehart & Co., New York, 1958, pp. 95–163.
2. Capeci, S.W., J.W. Osborn, A.J.W. Angell, and P. van Dijk, U.S. Patent 5,366,652 (1994).
3. Appel, P.W., P.L.J. Swinkels, and M. Waas, U.S. Patent 5,133,924 (1992).
4. Blackburn, S.N., F. Delwel, and E.H. Evans, U.S. Patent 4,826,632 (1989).
5. Lynn, J.L., Jr., in *Kirk-Othmer Encyclopedia of Chemical Technology,* 4th edn., edited by J.I. Kroschwitz, and M. Howe-Grant, Wiley, New York, 1993, Vol. 7, pp. 1103–1106.
6. Montague, P.G., and J.C. Van de Pas, U.S. Patent 5,147,576 (1992).
7. Bauer, H.E., M.G. Clarke, J.E. Lovas, W.R. Narath, and A.N. Williams, U.S. Patent 5,288,417 (1994).
8. Lang, D.J., A.A. Kamel, P.A. Hanna, R. Gabriel, and R. Thieler, U.S. Patent 5,200,236 (1993).
9. Gardlik, J.M., T. Trinh, T.J. Banks, and F. Benvegnu, U.S. Patent 5,102,564 (1992).
10. Jordan, N.W., W.E. Eccard, and J.R. Schwartz, U.S. Patent 5,393,449 (1995).

Chapter 5

Polymers for Detergents: Current Technology and Future Trends

William L. McCullen

  Rohm and Haas Co., Spring House, Pennsylvania

Polymers: Versatile Raw Materials

Over the last ten to fifteen years, the form and composition of household detergents have changed dramatically. Polymers have played a key role in this revolution, and the level of polymer usage has undergone dramatic growth. It is estimated that the worldwide consumption of polymers in 1993 exceeded 130,000 metric tons (1).

Polymers are most simply described as materials, either synthetic or naturally occurring, which have identifiable repeating units or monomers. In this respect, many surfactants are also polymers; however, I will restrict the scope of the discussion to materials which are *not* commonly considered to be surfactants. The exciting possibility offered by polymers is that they are often multifunctional; they can bring more than one benefit to the detergent formulation. Furthermore, polymers can be purposefully tailored to meet specific needs. Ideally, one can design a polymer to perform a specific function at a specific site in a multiphase system, the situation usually encountered in household cleaning applications. This can be accomplished by manipulating the following characteristics of a polymer:

- Composition: monomers can be selected based on functionality (anionic, cationic, and nonionic), size (relatively small or possessing bulky side chains), and solubility (hydrophilic *vs.* hydrophobic).

- Molecular Weight: polymer molecular weight can be varied from <1000 daltons to >1,000,000 daltons. To a large extent, this property determines the kinetics of the polymer solution phase reactions and polymer surface/interfacial adsorption characteristics.

- Morphology: polymers can be linear or branched. They can even be designed to form three-dimensional networks in solution through self-association as is observed for certain classes of acrylic-based rheology modifiers.

In short, the number of possible combinations is enormous. It takes little imagination to see how complex the interplay of these factors can be and how many possibilities exist to design polymers to provide specific benefits.

Traditional Applications of Polymers in Household Detergents

It is difficult to divide any history into a "before" and "after" or "the old" and "the new." Nonetheless, for the purpose of this perspective, I will claim poetic license and do just that. This section will describe polymer applications which I regard as well established and significant in the decades of the 70s and 80s. The next section will describe polymer applications which have become significant or appear to be increasingly important in this decade. This arbitrary distinction should not be construed to imply that nothing new and significant can come from the "old" technology nor should it be construed to mean that the "new" technology was conceived and developed entirely since 1990.

Builder Assists and Process Aids in Household Detergents: Polycarboxylates

As the use of sodium tripolyphosphate (STP) as the primary builder in household laundry detergents was reduced or eliminated in many regions of the world, new builder systems were explored. Despite many years of effort, including attempts to develop a cost-effective, biodegradable organic sequestrant, no single material has been identified which can replace the many benefits that STP brings to household laundry products (2), particularly hardness removal and soil dispersion. Instead, builder *systems* based on the inorganic components, zeolite A, sodium carbonate, and sodium silicate, emerged as the standard for premium detergents. Zeolite A was introduced to remove hardness from the wash bath via sodium-calcium ion exchange. Sodium carbonate and sodium silicate were incorporated to impart alkalinity while the latter brought corrosion inhibition benefits as well.

Fabric Incrustation. It was quickly realized that fabric washed in the new builder system soon developed a rough, faded appearance upon repeated washings, resulting from the deposition of zeolite and calcium carbonate on the fabric surface (3). The formation of insoluble calcium carbonate, due to the relatively slow rate of sodium-calcium ion exchange, coupled with poor dispersion of insoluble zeolite led to fabric incrustation. The problem was most severe in Europe where wash temperatures in the range of 60–90°C accelerated calcium carbonate precipitation.

Polycarboxylates (PCA), a class of materials which include homopolymers of acrylic acid (pAA) and copolymers of acrylic and maleic acid (AA/Mal), were found to function synergistically with zeolites to reduce fabric incrustation significantly (4). This is illustrated in Fig. 5.1 by a comparison of residual ash levels on cotton terry fabric laundered 10 times with the detergents shown in Table 5.1 under stressed European wash conditions: 6 g/L dosage, 600 ppm hardness as $CaCO_3$, 60°C. The Phosphate, Zeolite, and Zeolite/PCA bases are identical, with the exception of the levels of STP, zeolite, polymer, and sodium sulfate. The data show significant incrustation (1.3%) when zeolite is employed without polycarboxylates. Even the phosphate base exhibits slight incrustation which is probably due to the precipitation of insoluble calcium phosphate salts (3).

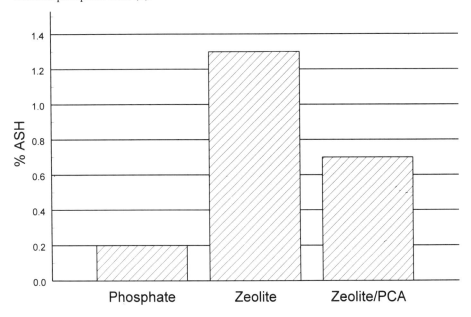

Fig. 5.1. Fabric incrustation: comparison of detergent bases in Table 5.1 under stressed European wash conditions (10 cycles). PCA, polycarboxylate.

TABLE 5.1 Detergent Bases—Fabric Incrustation Study in Figure 5.1[a]

Ingredient	Phosphate base (%)	Zeolite base (%)	Zeolite/PCA base (%)
STP	40	0	0
Zeolite A	0	40	31
AA/Mal (70K MW)	0	0	4.5
Sodium silicate	5	5	5
Sodium carbonate	15	15	15
Sodium sulfate	0	0	4.5
Sodium perborate	16	16	16
TAED	3	3	3
LAS	6	6	6
Alkyl ethoxylate	8	8	8
Soap	1	1	1
Minors	6	6	6

[a]Abbreviations: STP, sodium tripolyphosphate; AA/Mal, acrylate/maleate; TAED, tetraacetylethylenediamine; LAS, linear alkylbenzenesulfonate.

Polycarboxylates reduce incrustation primarily through two mechanisms (4,5). The first is crystal growth inhibition in which the polymer adsorbs on the growing crystals and interferes with the kinetics of particle growth. As a result, smaller crystals are produced which minimizes the likelihood of deposition. The second is particle dispersion in which polymer adsorption enhances the net negative surface charge of the particles which prevents interparticle flocculation and subsequent deposition. Polycarboxylates can also sequester hardness; however, at levels normally employed in detergents (<6%), the contribution to hardness removed by sequestration is not significant compared with that of the primary inorganic builders.

Soil Dispersion. Another major benefit that STP brings to the washing process is soil dispersion. Phosphates are recognized to function with surfactants to remove particulate soils from fabric surfaces as well as to suspend particulate soils in the wash bath (6). This occurs because STP adsorbs onto pigment soils, enhancing the negative surface charge of the particle and facilitating its removal from the fabric surface. As described in the previous section, polycarboxylates adsorb onto particle surfaces in an analogous manner and also function as soil dispersants. The soil removal data in Fig. 5.2 illustrate the soil dispersion benefits which polycarboxylates bring to zeolite-based detergents (Table 5.2) under typical U.S. wash conditions: 1.3 g/L dosage, 100 ppm hardness as $CaCO_3$, 40°C. For each fabric, the incorporation of 3% sodium polyacrylate into the formulation results in an increase in detergency of five units.

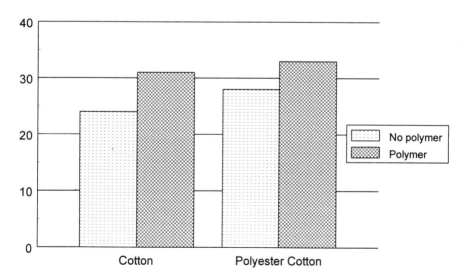

Fig. 5.2. Soil removal: effect of polycarboxylates in zeolite-built detergents under typical U.S. wash conditions.

TABLE 5.2 U.S. Compact Powdered Laundry Detergent Prototype[a]

Ingredient	(%)
Sodium carbonate	40
Zeolite A	30
Sodium silicate	5
LAS	7.5
Lauryl sulfate	14.5
pAA (MW 4500)	0–3.0

[a]*Abbreviations:* LAS, linear alkylbenzenesulfonate; pAA, sodium polyacrylate.

Process Aid in Detergent Production. In addition to performance benefits, polycarboxylates are also effective process aids in the production of detergents. In the traditional spray-drying process, polycarboxylates are employed as dispersants for crutcher slurries to enhance productivity by enabling higher solids throughput. Polycarboxylates have also been found to improve the physical integrity of the final detergent particles and to minimize the production of fines.

Antiredeposition Aids

Once soil is removed from the fabric, it must be suspended until the wash liquor is separated from the fabric. Most detergents, therefore, contain a small percentage of an antiredeposition agent (0.5–1.0%) to prevent fabric graying over multiple wash cycles due to soil deposition. There are literally hundreds of polymers claimed in the patent art as antiredeposition aids (ARD). The most commonly employed are carboxymethylcellulose for cotton, hydroxymethylpropylcellulose for synthetic fabrics (7), and polycarboxylates.

Emerging Polymer Applications and Future Needs

With the introduction of compact powdered detergents in 1989, we have witnessed major changes in household detergent technology. The primary drivers of this change are improved cost performance for the consumer and environmental concerns. While these drivers are not new, they have continued to spawn the development of new raw materials as well as new production process technology. These changes have been reviewed in the literature (8–10) and are beyond the scope of this paper. Instead, I will review several trends and initiatives which have and will continue to influence the use and development of polymers for laundry detergents.

Higher Density Detergents

The shift to compact detergents in the last few years has ushered in new granulation processes (10), such as agglomeration and extrusion, which, in combination with dry blending, have reduced or eliminated reliance upon the traditional spray-drying operation. This shift has placed new demands on polycarboxylates. First, the nature of agglomeration and dry blending requires that the moisture introduced into the process be minimized. Accordingly, water introduced by polycarboxylates (generally supplied as a 40–45% actives aqueous solution) can be a limiting factor in the use of these polymers. Second, detergents made via dry blending will not remain thoroughly mixed upon handling and transportation unless all of the particles have approximately the same particle size distribution and particle density. Traditionally, spray-dried polycarboxylates were low-density products (400 g/L) with a highly variable particle size distribution. Consequently, the polymers are not optimized to blend directly into the new detergents with densities in the range of 550–900 g/L (11,12).

One polycarboxylate supplier introduced a granular product with average particle sizes of 300–500 µm, a narrow particle size distribution, and reduced hygroscopicity. Unfortunately, the particle densities achieved through this advance did not match those of the higher-density compact detergents. Another supplier introduced a granular form of polycarboxylates prepared on a variety of inorganic carriers which are commonly used in detergents, such as sodium carbonate, sodium silicate, and zeolite A. These granules, which contain 30–40% polymer by weight, were prepared with densities in excess of 750 g/L and a particle size distribution closely matching that of typical powdered products (13).

New Builder Systems

Recently there has been much interest in new builder systems. A new layered silicate from Hoechst has been introduced into household laundry detergents by Procter & Gamble in Europe, and a granular form of sodium carbonate/sodium disilicate from Rhone-Poulenc is being promoted (14). There also appears to be renewed interest in the use of sodium citrate.

The new layered silicates and carbonate/silicate granules undoubtedly bring new benefits to the detergent formulation; however, both types of builders can lead to fabric incrustation when formulated without polycarboxylates. Cotton terry fabric was laundered with the detergents shown in Table 5.3 which include a no-polymer and polymer-containing variant of the zeolite, layered silicate, and carbonate/disilicate built bases. Figure 5.3 presents a comparison of residual ash levels found after 10 wash cycles under stressed European conditions: 6 g/L dosage, 600 ppm hardness as $CaCO_3$, 60°C. In all cases, the nonpolymer formulations exhibited significant incrustation which was reduced significantly with the incorporation of a modest level of acrylate/maleate copolymer (4.5%).

It seems unlikely that either of these new builders will replace zeolite A completely. Instead, a strategy to blend builders will continue to be the preferred path.

TABLE 5.3 Detergent Bases—Fabric Incrustation Study in Fig. 5.4[a]

Ingredient	Zeolite (%)	Zeolite polymer (%)	Layered silicate (%)	Layered silicate polymer (%)	Carbonate/ silicate granule (%)	Carbonate/ silicate granule polymer (%)
Zeolite A	40	31	0	0	0	0
Layered silicate	0	0	40	31	0	0
Carbonate/silicate granule	0	0	0	0	40	31
AA/Mal (70K MW)	0	4.5	0	4.5	0	4.5
Sodium silicate	5	5	5	5	5	5
Sodium carbonate	15	15	15	15	15	15
Sodium sulfate	0	4.5	0	4.5	0	4.5
Sodium perborate	16	16	16	16	16	16
TAED	3	3	3	3	3	3
LAS	6	6	6	6	6	6
Alkyl ethoxylate	8	8	8	8	8	8
Soap	1	1	1	1	1	1
Minors	6	6	6	6	6	6

[a]See Tables 5.1 and 5.2 for abbreviations.

How this will evolve is yet to be seen, but it is important to realize that, as builder blends change, the kinetics and mechanism of hardness removal also change which can affect the optimum polymer selection.

Unilever recently introduced a new zeolite, Zeolite MAP, into several of their formulations in Europe (15). This new zeolite, which is reported to have a different structure and crystal size, is said to offer improved calcium exchange properties. It was also reported that polycarboxylates were removed from formulations containing zeolite MAP. We must wait to see if this or other novel ion-exchange materials will diminish the need for scale inhibitors to handle the problem of fabric incrustation.

Color Care Detergents—Dye Transfer Inhibitors

Color-safe laundry detergents have emerged in the 1990s as a small but important segment of the household market, particularly in Europe (9). The benefit to the consumer is the preservation of fabric color without the requirement to sort laundry. A key technical requirement of a color-safe detergent formulation is prevention of the transfer of dyes among fabrics in the wash.

The most common commercial technology employed to inhibit dye transfer is polyvinylpyrrolidone (PVP), which has been used for years in the textile industry. It is believed that PVP functions by stoichiometric association with dye molecules in the wash bath, stabilizing them in solution and preventing subsequent deposition on fabric

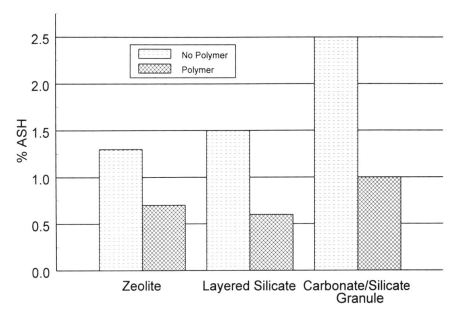

Fig. 5.3. Fabric incrustation: effect of polycarboxylate in the detergents shown in Table 5.3 under stressed European wash conditions (10 cycles).

(16, 17). Other polymers containing nitrogen heterocylic groups, such as polyvinylimidazole (16) and poly 4-vinylpyridine N-oxide (18), have also been identified as effective dye transfer inhibitors.

When used at fairly low levels in detergents (<0.5%), PVP is quite effective on certain classes of dyes, but the fact that its effectiveness is not universal is a clear limitation. The effectiveness of PVP (MW 36,000) was evaluated for six dyes, each representing a different dye class. White cotton linen was washed with a commercially available ultradetergent in a European front-loading machine under U.S. conditions: 0.7 g/L dosage, 150 ppm hardness as $CaCO_3$, 45°C, bath/fabric = 8/1. In each case, the dye was added directly to the wash bath; the reflectance was read after one wash cycle. The results in Fig. 5.4 compare the performance of PVP added at 75 ppm with the no-PVP control. PVP is clearly effective on the examples of direct, acid, reactive, and vat dyes shown here, but it fails to inhibit the transfer of the disperse and basic dyes. As consumer demands increase for more effective color-safe detergents, the opportunity exists to develop dye transfer inhibitors with a broader spectrum of activity or inhibitors that complement the activity of the currently available polymers.

Biodegradable Components

Like Western Europe, the U.S. has become an environmentally conscious society. This awareness continues to spark concern and questions about the long-term effects

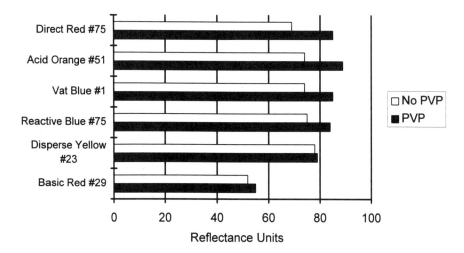

Fig. 5.4. Dye transfer inhibition: utility of polyvinylpyrrolidone (PVP) for various dyes.

of chemicals discharged into the environment. As a consequence, consumers and legislative bodies are calling for products which are more environmentally friendly. The polycarboxylates currently used in the detergent and cleaners industry have been studied extensively (19–21). While these products are not regarded as biodegradable, they have been found to be innocuous in the environment. In spite of this, the need exists to develop completely biodegradable, cost-effective replacements for the current polymers in anticipation of future consumer and legislative mandates.

In a discussion of this subject, it is imperative to define biodegradability. For the purpose of this paper, I will use the definition that a polymer is biodegradable if it can be shown to completely degrade to carbon dioxide, water, and mineral salts in its receiving environment within a reasonable time. Because most household waste water flows into municipal sewage treatment systems, a biodegradable polymer must be completely removed from the sewage treatment plant effluent and must degrade completely in an activated sludge environment.

Considerable effort has been made toward the development of completely biodegradable scale inhibitors and dispersants for household laundry detergents (22). From the perspectives of biodegradability and performance, the most promising technology to date is the development of sodium polyaspartate, p(Asp). When p(Asp) polymer is produced in a manner which yields a linear polymer, SCAS (Semi-Continuous Activated Sludge) and CO_2 evolution studies, which are recognized models of a sewage treatment system, reveal that p(Asp) meets the strict definition of biodegradability. The performance of p(Asp) also shows considerable promise. The results in Fig. 5.5 compare the performance of sodium polyaspartate and sodium polyacrylate for clay soil removal and whiteness maintenance on cotton when evaluated in the formulation of Table 5.2 under U.S. wash conditions (40°C, 100 ppm hardness as

CaCO$_3$, 1.3 g/L dosage). Both polymers performed equivalently, exhibiting benefits in detergency and soil antiredeposition.

Despite considerable interest in this technology, the cost of sodium polyaspartate has inhibited its broad adoption. The production of the completely biodegradable form of sodium polyaspartate relies on the use of L-aspartic acid as a starting raw material. At present, neither the scale of production nor the process costs of L-aspartic acid appears to be able to compete with acrylic acid and maleic anhydride. As a consequence, it seems as if the adoption of sodium polyaspartate must await progress in the economics of L-aspartic acid production.

Soil Release Polymers

The basic concept of soil release is to modify the fabric surface, making it easier to clean. A material that does this is commonly known as a soil release polymer (SRP). Once adsorbed at sufficient levels, the SRP functions by increasing the hydrophilicity of the fabric surface which enhances the wettability of the fabric surface and inhibits the penetration of oily soils. As a consequence, soils can be more easily removed. The most common soil release polymers used today are based on polyethyleneterephthalate (PET) which may be modified for compatibility and performance in detergents, fabric softeners, and dryer-added sheets. Typical modifications include copolymerization with polyethylene glycol (PEG) (23) and/or end-capping with sulfonates (24).

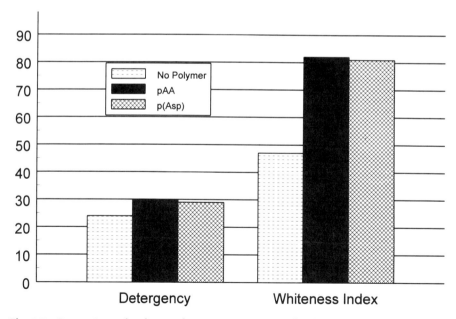

Fig. 5.5. Comparison of sodium polyaspartate, p(Asp), and sodium polyacrylate, pAA, in a zeolite-built detergent under typical U.S. wash conditions.

Modified cellulosics (25) and alkylene oxide/vinyl ester copolymers (26, 27) are also claimed to function as soil release agents.

The SRPs described above are primarily effective on hydrophobic fabrics such as polyester. Unfortunately, they are much less effective on polyester/cotton blends and almost completely ineffective on cotton. Two commercially available SRPs, a poly(ethylene oxide/vinyl acetate) [p(EO/VA)] copolymer and a PET-based product, were evaluated for effectiveness on the removal of used motor oil stains from polyester, a polyester/cotton blend (PE/C), and cotton. The soiled fabrics were washed in a front-loader machine with a commercially available liquid laundry detergent under the following conditions: 1.9 g/L dosage, 190 ppm hardness as $CaCO_3$, 40°C, bath/fabric ratio = 8/1. The results presented in Fig. 5.6 demonstrate the benefits of the SRPs on polyester, particularly with the PET-based product under these conditions. The beneficial effects, however, are substantially reduced for polyester/cotton; there is no performance benefit observed on cotton. Because most wearable consumer fabrics are cotton or cotton blends, this deficiency severely limits the use of current SRPs and presents an interesting and challenging objective for the future.

Nonphosphate Automatic Dishwashing Detergents

The use of phosphates in household laundry detergents has been regulated or eliminated for years in many parts of the world, but phosphate use in automatic dishwashing

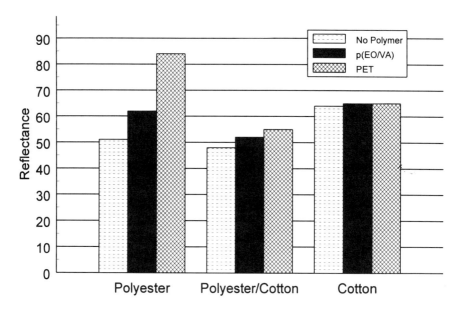

Fig. 5.6. Performance of soil release products under U.S. wash conditions. PET is a polyethyleneterephthalate-based product; p(EO/VA) is a copolymer of ethylene oxide and vinyl acetate.

detergents (ADD) remains largely unregulated. Nonetheless, the major manufacturers of ADD in Europe now offer nonphosphate-containing brands, and it is reported that Procter & Gamble introduced a nonphosphate version in a U.S. test market in late 1994 (28).

Somewhat analogous to the problem of fabric incrustation observed with nonphosphate laundry detergents, the problem of filming on glassware due to the deposition of inorganic salts becomes a performance concern for nonphosphate ADD formulations. Because the use of zeolites in ADD seems unlikely at present, it appears that the nonphosphate builder systems will be combinations of sodium citrate (a soluble sequestrant), sodium carbonate/bicarbonate, and sodium silicates.

As one might predict, the filming performance of a nonphosphate detergent is heavily dependent upon the balance of precipitating and sequestering builders. This effect is shown by comparing the glassware filming performance of the ADD formulations presented in Table 5.4. All of the formulations contain the same components, but the formulations labeled "carbonate" are weighted toward higher levels of precipitating builders, whereas the formulations labeled "citrate" are weighted toward higher levels of sequestering builders.

The data in Fig. 5.7 compare the filming performance of these detergents on a scale of 0 (no film) to 4 (heavy film) under the following conditions: standard U.S. dishwasher, 5 wash-dry cycles, 27 g/wash dosage, no food soil, 300 ppm hardness as $CaCO_3$, 50°C. In addition, the filming performance of a premium U.S. phosphate-based detergent (P Base) is presented as a benchmark. The less expensive carbonate formulation is completely unacceptable, whereas the performance of the more expensive citrate formulation is equivalent to that of the phosphate-based product. The addition of sodium polyacrylate at modest levels (6%) significantly improves the filming performance of the carbonate formulation. Clearly, there is an opportunity to use poly-

TABLE 5.4 Nonphosphate Automatic Dishwashing Formulations[a]

Ingredient	Carbonate (%)	Carbonate with polymer (%)	Citrate (%)
Sodium carbonate	30	30	10
Sodium bicarbonate	20	20	10
Sodium citrate	10	10	30
Sodium disilicate	7	7	10
EO/PO surfactant	3	3	3
Sodium perborate	7.5	7.5	7.5
TAED	2.5	2.5	2.5
Protease	1	1	1
Amylase	1	1	1
Polymer	0	6	0
Sodium sulfate	18	12	25

[a]*Abbreviations:* EO/PO, ethylene oxide/propylene oxide; TAED, tetraacetylethylenediamine.

carboxylates as a key ingredient in the formulation of a cost-effective nonphosphate ADD. Furthermore, it is likely that further study will yield more efficient polymers than those developed primarily for household laundry use.

New U.S. Washing Machine Standards

It is expected that the U.S. Department of Energy will issue standards in 1997 or 1998 which will take effect in the year 2000 regulating the energy consumption of washing machines. The likely consequence will be new machine designs that will reduce the use of hot water. If this does occur, three things seem relatively certain:

- Less water will be consumed per load of clothes.
- The average water temperatures will decrease.
- Current detergents will have to be reformulated but usage will decrease minimally, if at all, because dosage is more a function of soil load than water consumption.

What will be the impact on polymer usage in detergents? Surely, it is impossible to predict; however, I believe that polymers will continue to be a key ingredient in household laundry formulation strategies. With less water in the wash of the future, the benefits polymers bring, such as soil dispersion and antiredeposition, become

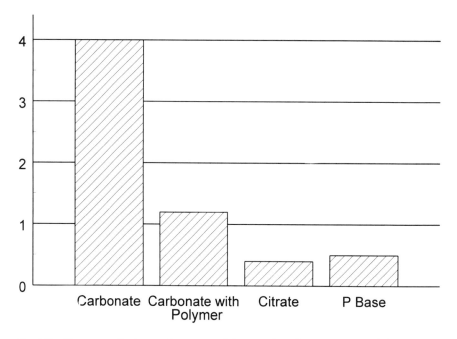

Fig. 5.7. Glassware filming performance of the nonphosphate automatic dishwashing detergent (ADD) formulations shown in Table 5.4. P Base is a commercial phosphate-based ADD.

increasingly important. In addition, despite lower wash temperatures which favor solubilization of calcium carbonate, fabric incrustation is likely to remain a performance issue because the levels of carbonate/silicate alkalinity in the wash of the future will increase along with overall detergent concentrations. Certainly, the multifunctional benefits that polymers bring to detergents will become *more* important as the wash of the future becomes more dependent on chemical energy and less dependent on thermal energy.

Summary

The need to provide consumers with more cost-effective, environmentally friendly products will drive the development of household detergent technology. New raw materials must be efficient, multifunctional, and safe. In addition, the cycle time for the development of new raw material must decrease because product life cycles of consumer products are growing shorter.

Detergent polymers have played a key role in the evolution of detergents over the past fifteen years, and many untapped opportunities exist. The key to future success will be the ability of polymer suppliers and detergent manufacturers to work jointly to capitalize on the unique technical capabilities of each in order to bring new products to the marketplace quickly.

Acknowledgments

I want to thank my colleagues at Rohm and Haas Company and NorsoHaas S.A. for their cooperation in gathering technical data. I also wish to recognize Dr. R.G. Aviles, Dr. M.B. Freeman, Dr. C. Schwartz, and Dr. J.E. Shulman for their help in writing and editing this paper.

References

1. Perner, J., Detergent Polymers, in *Proceedings of the 3rd World Conference on Detergents: Global Perspectives,* edited by A. Cahn, AOCS Press, Champaign, 1994, pp. 168–173.
2. Cahn, A., Builder Systems in Detergent Formulations, *INFORM* 5:70–74 (1994).
3. Diessel, P., J. Stabenow, and W. Trieselt, Mode of Action of Polycarboxylates in Detergents: The Mode of Action of Polycarboxylates in Phosphate-Reduced and Phosphate-Free Detergents, As Seen with the Electron Microscope, *Tenside Surfact. Det.* 25:268–274 (1988).
4. Andree, H., P. Krings, H. Upadek, and H. Verbeek, Possibilities of Combining Zeolite A with Different Cobuilders, *Second World Conference on Detergents,* edited by A.R. Baldwin, 1987, pp. 7–11.
5. Zini, P., *Polymeric Additives for High Performing Detergents,* Technomic Publishing Company, Inc., Lancaster, Pennsylvania, 1995, pp. 30–65.
6. Jakobi, G., and A. Lohr, *Detergents and Textile Washing,* VCH Verlagsgesellschaft mbH, Weinheim, Germany, 1987, pp. 21–29.

7. Greminger, G.K., Jr., A.S. Teot, and N. Sarkar, Antiredeposition Additives Face New Opportunities, *J. Am. Oil Chem. Soc. 55*:122–127 (1978).
8. Grime, J.K., Laundry Technology Trends in the Americas, in *Proceedings of the 3rd World Conference on Detergents: Global Perspectives,* edited by A. Cahn, AOCS Press, Champaign, 1994, pp. 64–70.
9. Lee, A.E., Technology Developments in Laundry Products: Powders/Liquids in Europe, in *Proceedings of the 3rd World Conference on Detergents: Global Perspectives,* edited by A. Cahn, AOCS Press, Champaign, 1994, pp. 64–70.
10. Suzuki, A., Laundry Detergents in Asian and Pacific Countries, in *Proceedings of the 3rd World Conference on Detergents: Global Perspectives,* edited by A. Cahn, AOCS Press, Champaign, 1994, pp. 75–81.
11. *Agglomerations, 6(2),* Colin A. Houston & Associates, Inc., Mamaroneck, New York, (1995), pp. 1–5.
12. *Agglomerations, 6(3),* Colin A. Houston & Associates, Inc., Mamaroneck, New York, (1995), pp. 6–9.
13. Pilidis, A., Novel High Density Dry Polymers for Detergent Applications, in *Proceedings of the 3rd World Conference on Detergents: Global Perspectives,* edited by A. Cahn, AOCS Press, Champaign, 1994, pp. 251–252.
14. Milmo, S., A New Generation, *Chemical Marketing Reporter,* Schnell Publishing Company, New York, Jan. 16, 1995.
15. *Agglomerations, 5(2),* Colin A. Houston & Associates, Inc., Mamaroneck, New York, (1994), p. 2.
16. Jaeger, H., and W. Denzinger, Mode of Action of Polymers with Dye Transfer Inhibiting Properties, *Tenside Surfact. Det. 28*:428–433 (1991).
17. Hornby, J., PVP: A Dye Transfer Inhibitor for Laundry Detergents, *HAPPI 32*:88–89 (1995).
18. Convents, A.C., and A. Busch, European Patent 635563 A1 (1993).
19. Freeman, M.B., and T.M. Bender, An Environmental Fate and Safety Assessment for a Low Molecular Weight Polyacrylate Detergent Additive, *Environ. Toxicol. 14*:101–112 (1992).
20. Opgenorth, H., Polymeric Materials Polycarboxylates, in *The Handbook of Environmental Chemistry, 3(F),* edited by O. Hutzinger, Springer-Verlag, Berlin, 1992.
21. Chiaudani, G., and P. Poltronieri, Study of the Environmental Compatibility of Polycarboxylates Used in Detergent Formulations, *Ingegneria Ambientale 11*:1–43 (1990).
22. Paik, Y., E. Simon, and G. Swift, A Review of Synthesis Approaches to Biodegradable Polymeric Carboxylic Acids for Detergent Applications, *ACS Series—Advances in Chemistry No. 248,* in press.
23. Ross, A.M., and D.F. Kirkwood, U.S. Patent 4,795,584 (1989).
24. Gosselink, E.P., U.S. Patent 4,721,580 (1988).
25. Burns, M.E., U.S. Patent 4,174,305 (1979).
26. Holland, R.J., O.K. Bullard, A.V. York, D. Boeckh, W. Trieselt, P. Diessel, and H. Jaeger, U.S. Patent 5,049,302 (1991).
27. Boskamp, J.V., European Patent Application 358472 A2 (1989).
28. *Agglomerations, 5(6),* Colin A. Houston & Associates, Inc., Mamaroneck, New York, (1994), p. 4.

Chapter 6

Detergent Enzymes—Global Trends

Hans A. Hagen

  Novo Nordisk, Denmark

Introduction

Ten years ago, if you looked for enzymes in household detergent products, you would find few other than proteases, and of those, you would find only three or four different kinds. Today, the picture is strikingly different: in many leading brands world-wide, you will find a mix of two, three, or even more different enzymes, each designed to offer its own specific contribution to the cleaning process. In almost every case, however, the mix will contain a protease, and more than ten different proteases are available to the detergent industry market today. Within the group of "novel" detergent enzyme classes, e.g., amylases, lipases, and cellulases, the variety of commercially available products is still much smaller, but it is growing at an impressive rate. (See Table 6.1.)

What has happened—and what is happening? I believe most of us know the answer to that question: the driving forces behind the growth in detergent enzyme variety are found both in the marketplace and in the field of biotechnology.

TABLE 6.1 Examples of Commercial Detergent Enzyme Products (1995)

Protease	Amylase	Lipase	Cellulase
Alcalase[a] Durazym[a]	Amylase T[c]	Lipolase[a]	Carezyme[a]
Esperase[a] Maxacal[b]	BAN[a]	Lipolase Ultra[a]	Celluzyme[a]
Maxapem[b] Maxatase[b]	Duramyl[a]	Lipomax[b]	Cellulase EG 3[d]
Opticlean[c] Purafect[d]	Maxamyl[b]	Lumafast[d]	KAC-5000 (3)[e]
Purafect OxP[d] Savinase[a]	Termamyl[a]		

[a]*Sources:* Novo Nordisk; [b]Gist Brocades; [c]Solvay; [d]Genencor International; [e]Kao Corporation.

Market Factors Driving Detergent Enzyme Development

Lower Washing Temperature

Unlike protein stains, starch and fat stains are relatively easy to remove when washing in hot water. Consequently, the commercial demand for detergent amylase and lipase has reached its current high level mainly as a result of the gradual decrease in household laundry and dishwashing temperature. Many detergent brands now contain three or four different enzymes, and to facilitate the handling and dosing in the detergent plant, pre-blends of enzymes such as protease, lipase, and amylase are available from some enzyme suppliers.

More Compact Formulations

The compact detergent concept, which is derived mainly from an environmental focus, has increased the importance of one of the classical virtues of enzymes, namely, their excellent space efficiency, which is explained mainly by their catalytic nature. In recent years, the space efficiency of detergent enzymes has been improved further through stepwise increases in the concentration of enzyme protein in many commercial protease and amylase products, in both granular and liquid form. Today, leading detergent manufacturers are able to dose protease granulates that are 3–4 times stronger than traditional "single-strength" products. Such high-strength granulates are economically attractive because of the relatively high cost contribution of granulation and encapsulation.

More Diverse Detergents

Market factors behind the current large-scale application of detergent cellulase include the success of the "casual cotton" garment fashion and the increasing market share of detergent brands specifically developed for the washing of colored garments. The increasing product diversity in the household detergent market has also resulted in the development of specialty proteases for bleach-containing powders and low-to-medium alkaline liquids.

More Focus on the Environment

The steadily increasing focus on protection of the environment has resulted in many detergent ingredient substitutions in both the surfactant and builder fields. It is nearly impossible, though, to find any type of active cleaning ingredient that approaches the environmentally benign nature of enzymes. Apart from their excellent space and weight efficiency, enzymes are extremely easily degraded in nature, and sometimes even exert a positive effect on activated sludge in biological sewage treatment plants.

If we look at the field of automatic dishwashing detergents (ADDs), environmental considerations are not the sole influence on enzyme use. In this market, detergent compositions are also changing with a view toward consumer safety, resulting in the replacement of harsh chemicals such as strong alkali- and chlorine-based bleaching systems with milder ingredients. Here, enzymes are among the tools available to ensure that the cleaning power of the new milder detergent formulation is up to standard; protease and amylase are being used increasingly in these second-generation ADDs.

Biotechnology Factors Driving Detergent Enzyme Development

An example of genetic engineering utilized widely in commercial production of detergent enzymes is the generation of "multicopy" production organisms. By introducing additional copies of a gene coding for a desired enzyme into a microorganism, the efficiency of the fermentation process in question can be improved; as a consequence, cost savings can be obtained which may be crucial for the survival of a given enzyme in the highly competitive marketplace. Examples include the proteases Savinase and Durazym, and the amylase Termamyl.

Even more exciting prospects are seen in the technology of transferring enzyme coding genes from one species to another. Through classical screening of microorganisms found in the wild, a number of interesting enzymes of potential use as detergent ingredients have been identified through the years. The growing demand for novelty or improved enzyme properties has forced researchers to look into more and more exotic species. In an increasing number of cases, these species have proved to be very difficult to ferment on an industrial scale or have been found to produce unwanted or even harmful substances in addition to the desired enzyme. As a consequence, such attractive new-found enzymes could not be produced on a large scale, as necessary to make their use as detergent ingredients feasible.

The development of genetic engineering technology has changed this situation dramatically. Now, the gene coding for a new attractive enzyme can be moved over from an exotic and difficult "wild" donor species into a so-called "host organism" which can be selected for its safety and ability to perform well in industrial-scale fermentation. The first commercial detergent lipase (Lipolase) was made available through this technology in 1988; others have followed, including the lipase Lumafast, the protease Purafect, and the cellulase Carezyme.

Protein engineering is the term most often used for the technology of improving enzyme performance through modification of the enzyme molecule. The modification is typically obtained by replacing a few specific amino acids in the enzyme protein structure by other amino acids with different properties. In this way, the enzyme's sensitivity to oxidation agents, alkali, ionic strength, temperature, or other relevant application parameters may be changed. The protein engineering tool has already been used to create a number of commercially available detergent enzymes, including the proteases Durazym (improved bleach tolerance), Maxapem, and Purafect OxP; the lipases Lipolase Ultra (improved stain affinity) and Lipomax CXT; and the amylase Duramyl (improved bleach tolerance).

Future Developments in Enzymes

What will happen in the future? No doubt, the enzyme industry will strive to utilize the powerful tools of biotechnology to improve the properties of current products and to develop novel enzymes that meet the needs of the detergent manufacturer.

Interface Affinity

In both laundry and dishwashing, enzymes are expected to be active on the interface between the washing liquor and the soil substrate. This means that to optimize the bottom-line efficiency of enzymes, one must try to maximize the affinity of the enzyme molecules towards this interface and minimize the amount of enzyme distributed randomly in the washing liquor. Selective modification of the enzyme molecule's electrical charge profile through protein engineering is one way of pursuing this goal. This concept has in fact been used in the development of some new lipases and proteases already available in the market.

Compatibility

A modern washing liquor or a detergent package is not a natural place for any enzyme to reside. Consequently, it is no wonder that many enzymes found in nature are not sufficiently compatible with standard detergent ingredients to be useful to the industry. Again, protein engineering may be an efficient tool to modify enzyme molecules in a way that makes them more tolerant to ingredients such as surfactants, builders, and alkali.

Low-Temperature Efficacy

Although detergent enzymes are often mentioned in connection with energy saving and lowered washing temperature, they do lose some of their cost-performance benefits in actual cold-water washing in the 10–20°C range. The mechanism behind this loss of activity is not fully understood, but there is some expectation that enzyme molecules may be modified through protein engineering in such a way that they retain more of their cleaning performance at very low washing temperatures.

In addition to these rather general ideas, it should be possible to give some more concrete examples of detergent enzyme novelties that are likely to appear in the near future.

Lipase

The cleaning effect of detergent lipases available today is seen mainly after two or more wash cycles, unless the laundry is subjected to prespotting or presoaking. This is because the enzymatic breakdown of the fat takes place at a sufficiently high rate only at the reduced moisture levels occurring during laundry drying. It is an obvious objective for protein engineering to improve the performance of lipases at the conditions found in washing systems in such a way that the enzyme effect is significant after only a single short-cycle wash.

Cellulase

To prevent unwanted fiber weakening, detergent cellulases with softening and color-clarification effects on cotton must be dosed with moderation. This limits the speed at which their beneficial effects on the laundry can be obtained. The use of sophisticated biotechnology tools including protein engineering is likely to be a way of developing fabric softening and color-clarification cellulase variants which can be safely dosed at higher levels.

Amylase

In automatic dishwashing, starch removal is a problem with second generation, mild detergents unless an amylase is incorporated. On the other hand, oxidative bleaching systems also seem to be indispensable, and they typically reduce the cost-performance benefits of the amylase. As a consequence, the interest in bleach-stable amylases, such as the very recently introduced Duramyl, is quite high, and again expectations focus on protein engineering as a way to attain the goal.

Enzymes for Bleaching

Different groups of enzymes that catalyze oxidation reactions have been known for many years, and their potential application in detergent bleaching systems has been the subject of much research. The objective would be an enzyme system that could ensure economic bleaching at low washing temperature with a minimum of harm to fabrics and environment. The research has focused mainly on the following:

- Oxidases which catalyze the reaction between molecular oxygen and a specific substrate molecule under formation of hydrogen peroxide.
- Laccases which catalyze the reaction between molecular oxygen and a specific substrate molecule under formation of water.
- Peroxidases that catalyze oxidation reactions with hydrogen peroxide.

Detergent bleaching systems based on oxidase have a weakness which may explain why they have never been commercialized: the amount of substrate chemical needed for the enzymatic reaction to form a sufficient quantity of hydrogen peroxide tends to be too large to be acceptable in modern concentrated detergent products. This weakness is not seen with peroxidase. For peroxidase, however, the challenges of cost efficiency and system control are considerable. With the strong biotechnology tools available today, it would not be surprising though to see peroxidase-based bleaching or dye transfer inhibition systems come to the detergent marketplace within a few years.

Biotechnology is Not the Only Thing

The demands on any detergent enzyme product include quite a few quality parameters that have little to do with biotechnology. Because of the dynamic developments in detergent product formulation taking place, the demands on enzyme contributions tend to broaden. If we look at granulated enzyme products, the dust and frangibility parameters have always been very much in focus due to their safety implications. Fortunately, enzyme producers and detergent manufacturers have been working successfully together to bring this concern under control on an almost global basis; I am confident that the same success will be achieved in those parts of the world where production and application of detergent enzymes are still in their infancy.

As already mentioned, the activity level of detergent enzymes used in products has been increasing, mainly as a result of the industry's need for controlling ingredient cost. For the enzyme producer, this development makes necessary a combination of process adjustments with a view to maintaining or improving other enzyme characteristics such as homogeneity, color, or odor. Typically, the non-enzyme biomass from the fermentation process is being removed from commercial enzymes to a greater and greater extent.

One problematic effect of the increasing purification of the enzyme fermentation broth is the removal of biomass components that have a stabilizing effect on the enzyme's molecular structure and thereby contribute to the storage stability of the final enzymatic detergent. As mentioned above, protein engineering is obviously one way of improving enzyme stability, but incorporation of selected stabilizing compounds, or use of protective granulate coatings may be important alternatives.

As a result of the increasing use of enzyme combinations in many detergent products, much effort is being spent in solving the problem of interenzyme compatibility, especially in liquid detergents. This is another area in which biotechnology as well as product formulation skills will be used in concert. It is also a prime example of the importance of the detergent formulator when it comes to optimizing enzyme performance. A major challenge for both enzyme producer and detergent manufacturer is found here: An enzyme product that has demonstrated excellent performance in one detergent may not show at all the same efficiency in another formulation. Although understanding of the interaction between enzymes and other detergent components is steadily growing, much empirical trial and error testing is still required to attain optimum benefit from the combined ingredients.

Biotechnology and other enzyme development tools are obviously only one part of the process. Successful application of detergent enzymes also depends on the ability of the detergent producer to formulate the product in a way that ensures a high degree of enzyme compatibility.

Chapter 7

On Understanding Hydrophobicity

Bengt Kronberg[a], Miguel Costas[b], and Rebecca Silveston[c]

[a]Institute for Surface Chemistry, S-114 86 Stockholm, Sweden, [b]Departamento de Fisica y Quimica Teorica, Facultad de Quimica, Universidad Nacional Autonoma de Mexico, Mexico D.F. 04510, Mexico, and [d]INRS-Energie et Materiaux A/S IMI/CNRC, Boucherville, QC, Canada J4B 6Y4

Introduction

In the micellization process, the hydrocarbon moiety of a surfactant molecule is transported from an aqueous environment into a hydrocarbon environment. This process is driven by the poor compatibility of hydrocarbons and water. Similarly, in the adsorption process of surfactants onto a solid/water interface or on the air/water surface, the hydrocarbon moiety is transferred from water to a hydrocarbon environment. Finally, the solubility of liquid hydrocarbons in water can be treated thermodynamically in terms of a transfer from the aqueous phase to the pure hydrocarbon liquid phase. This chapter concerns the thermodynamic analysis of micellization and adsorption of surfactants. Solubility data of hydrocarbons in water provide the basic source of information.

The Two Driving Forces for the Hydrophobic Effect

Previous work has shown that two independent effects contribute to the solution characteristics of hydrocarbons and surfactants in water (1–7). The first effect stems from the large amount of energy required to form a cavity in the water to accommodate the hydrocarbon moiety. This energy is regained upon micellization of a surfactant system, or upon phase separation of a hydrocarbon solution in water. In both of these processes, the hydrocarbon moiety is transferred from water to a hydrocarbon environment. This driving force therefore acts to promote micellization, adsorption, or phase separation. The origin of this effect is attributed to the high cohesive energy density in water, originating from intermolecular hydrogen bonds. Water has an exceptionally high hydrogen bond density due to the small size of the water molecules. For this reason, many hydrogen bonds have to be broken to create a cavity for the hydrophobe. In our previous analyses (4–7), this effect was considered to be purely energetic, i.e., there are no entropic parts. Thus, the enthalpy required for breaking a hydrogen bond is considered to be temperature independent.

The second effect counteracts the first one. It stems from the water that is structured in the immediate vicinity of a hydrophobe. Upon micellization, adsorption, or phase separation, this structured water is released, leading to a positive contribution to the entropy of micellization. It has been shown that this break up of ordered water molecules has to be accompanied by a positive enthalpy as well as a positive Gibbs free energy (8). This contribution therefore acts to prevent micellization, adsorption of surfactants, or phase separation of hydrocarbons.

The temperature dependence of the thermodynamic functions for the micellization, adsorption, and phase separation processes is shown schematically in Fig. 7.1A–C. Here it is clearly seen that the enthalpic part of the water-structuring contribution, ΔH_s, and the cavity contribution, ΔH_o, have opposite signs. Their sum will be zero at around 20°C, as seen in Fig. 1C, leading to a minimum in the critical micelle concentration (CMC) or the solubility, or to a maximum in the adsorption (see below). We also note that the Gibbs free energy originating from the water structuring, ΔG_s, and the contribution originating from the cavity formation, ΔG_o, are of opposite sign, indicative of two contributions that oppose each other in the micellization process.

The present analysis was first postulated by Shinoda (1,2). It is in contrast to ordinary analyses normally found in textbooks, in which the water-structuring contribution is thought to lower the critical micelle concentration, or solubility, i.e., normally it is postulated that it is the water structuring that is the cause of the hydrophobic effect.

Micellization

The temperature dependence of the CMC of ionic surfactants in water is quite analogous to the temperature dependence of the solubility of hydrocarbons in water. Both systems display a pronounced minimum in the temperature range 15–25°C. Figure 7.2 shows the temperature dependence of dodecylpyridinium bromide (DPBr), as

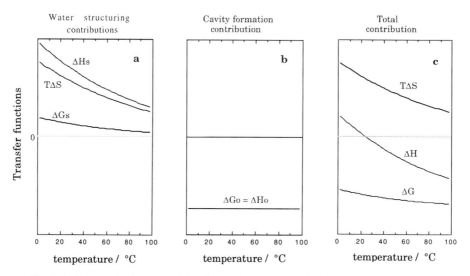

Fig. 7.1. Schematic drawing of the thermodynamic transfer functions for the transfer of a hydrophobic part of a molecule from water into a hydrophobic environment.

obtained from Ref. (9). This minimum is the result of the two opposing driving forces discussed above. The "normal" behavior would be that the solubility (or CMC) continuously decreases as the temperature is lowered. The unexpected increase in solubility (or CMC) at temperatures below 15–20°C is thus a result of the water-structuring contribution becoming increasingly important as the temperature is lowered.

Because the micellization process is cooperative, it can be treated as a phase separation, i.e., at the critical micelle concentration (CMC), the system is considered to separate into a micellar phase which is in equilibrium with the aqueous phase. Using such a phase separation model, the Gibbs free energy for the micellization process, ΔG_{mic}, is normally written as:

$$\Delta G_{mic} = RT \ln(CMC) \qquad [7.1]$$

where R and T are the gas constant and absolute temperature, respectively. It should be noted that the choice of concentration variable affects the ΔG_{mic} value obtained. In thermodynamic analyses, the mole fraction is often used as a concentration variable. Strictly, the CMC in Equation 7.1 should represent the activity of the surfactant at the critical micellar concentration. Using only the concentration in Equation 7.1 is equivalent to assuming that the activity coefficient is equal to one. In a more thorough analysis of the micellization process, an account of the activity, or activity coefficient, in Equation 7.1 has to be made. This activity coefficient is a reflection of the interaction

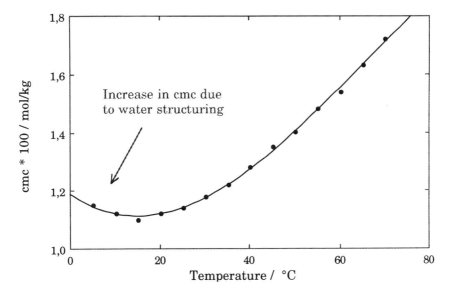

Fig. 7.2. The temperature dependence of the critical micelle concentration of DPBr in water [from Ref. (9)].

between the surfactant and the water. It is also, however, a reflection of the size difference between water and surfactant molecules, giving rise to a combinatorial entropy contribution to the Gibbs free energy of micellization in Equation 7.1. In evaluating the thermodynamics of micellization, it is important to take into account such contributions. This can be achieved by using the Flory-Huggins expression for the combinatorial entropy (4). For the Gibbs free energy of micellization we then have the following expression:

$$\Delta G_{mic} = RT \left(\ln X_{cmc} + \ln \frac{V_{hc}}{V_w} - \frac{V_{hc}}{V_w} + 1 \right) \quad [7.2]$$

where X_{cmc} is the critical micelle concentration expressed in mole fractions, and V_{hc} and V_w are the molar volumes of the hydrocarbon moiety and water, respectively.

The temperature dependence of ΔG_{mic} gives us both the enthalpy, ΔH_{mic}, and the entropy, ΔS_{mic}, of micellization, *viz.*,

$$\Delta H_{mic} = -T^2 d(\Delta G_{mic})/dT \quad [7.3]$$

and

$$\Delta S_{mic} = -\frac{d\Delta G_{mic}}{dT} = \frac{\Delta H_{mic} - \Delta G_{mic}}{T} \quad [7.4]$$

In Equations 7.3 and 7.4 we can assume as a good approximation that the molar volumes in Equation 7.2 are temperature independent. Reference (5) gives detailed information on how to obtain relevant thermodynamic information from the temperature dependence of the critical micelle concentration. We will not repeat the procedure but give only the results from such analyses.

Figure 7.3A shows the thermodynamic functions for the micellization of DPBr. We note that the enthalpy of micellization changes sign at the temperature at which a minimum of the CMC is found. Figure 7.3B shows the different contributions to the thermodynamic functions. Here subscript "*o*" refers to the contribution from the cavity formation and subscript "*s*" refers to the contribution from the structuring of water around the hydrophobe. First, we note that there is no entropy contribution from the cavity formation because the Gibbs free energy is assumed to be temperature independent. Second, we note that the two different contributions to the Gibbs free energy are of opposite sign and that the contribution from the water structuring is decreasing in magnitude with temperature. Third, we note that there is an enthalpy entropy compensation with respect to the water-structuring contribution, i.e., the enthalpy and entropy (times temperature) are of the same order of magnitude, whereas the Gibbs free energy is much smaller. Such behavior is typical for systems in which order (in our case, water structure) is formed or broken (8). According to the figure there will be no water-structuring contribution at temperatures above ~160°C.

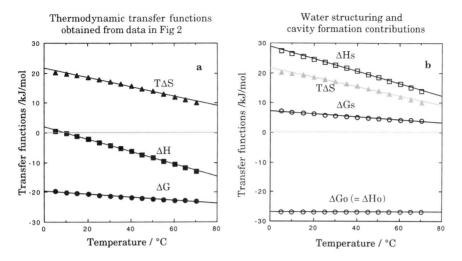

Fig. 7.3. The thermodynamic transfer functions for the micellization of DPBr.

Adsorption

The thermodynamic analysis for the adsorption process of surfactants is somewhat more complicated compared with that of the micellization process. Taking into account the difference in size between the surfactant and the water molecules, the following equation describes the partitioning of the surfactant between the surface (s) and the bulk solution (l) phase:

$$\ln\left(\frac{\phi_2^s}{(1-\phi_2^s)^r}\right) - \ln(\phi_2^l) = r\chi^l - r\chi^s(1 - 2\phi_2^s) + \frac{qa_1}{RT}(\gamma_1 - \gamma_2) \qquad [7.5]$$

where ϕ_2^l and ϕ_2^s are the surfactant volume fractions in the solution and the surface phase, respectively (10), and γ_1 and γ_2 are the interfacial free energies, or interfacial tensions, of the surface-water and surface-hydrocarbon part of the surfactant, respectively. The quantity a_1 is the cross-sectional area of an adsorbed water molecule, such that qa_1 is the cross-sectional area of the adsorbed segments of the surfactant molecule. The χ^l and χ^s parameters are the surfactant-solvent interaction parameters in the solution and surface phase, respectively. The equation has been used previously (10,11) in the analysis of surfactant adsorption on latexes of different polarity. It was shown that the surface polarity plays only a minor role (at most, ~20%) and that the hydrophobicity of the surfactants is the dominating (80–100%) force in the adsorption. Hence, even for the adsorption of surfactants on solid surfaces, hydrophobicity is at play, and hence the adsorption can be understood with the same terms as for micellization or phase separation.

Thus, for the micellization process, the adsorption of surfactants is split up into two contributions, i.e., one due to the high cohesive energy density of water and one due to the structuring of water around a hydrophobe. These two contributions counteract each other, thus causing a maximum in adsorption at around 20°C, which deKeizer *et al.* (12) demonstrated experimentally for the adsorption of the ionic surfactant dodecyl pyrridinium chloride (DPC) onto kaolinite. This maximum is in the same temperature range as that at which the surfactants show a minimum in the CMC. The temperature dependence of the adsorption of DPC on kaolinite is shown in Fig. 7.4. The normal temperature dependence of adsorption from solution is an increase in the adsorbed amount as the temperature is decreased. The decrease in adsorption at temperatures below ~20°C observed in Fig. 7.4 is thus a result of the water-structuring contribution becoming increasingly important. The thermodynamic functions for the adsorption process exhibit exactly the same picture as shown in Figure 7.3, and thus we conclude that the underlying thermodynamic behavior in the adsorption process is the same as in the micellization process.

In conclusion, we note that the water-structuring contribution to the hydrophobic effect is unfavorable in the sense that it counteracts hydrophobicity, whereas the contribution stemming from the large energy to form a cavity for the hydrophobe is favorable in the sense that it enhances hydrophobicity. Indeed, it is this latter contribution that is the cause for the hydrophobicity of surfactants and other molecules with hydrophobic groups.

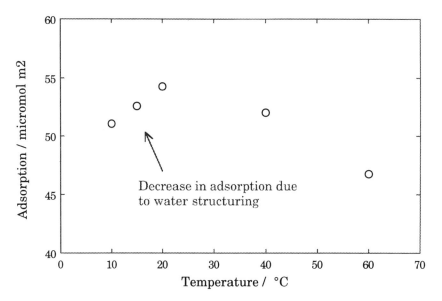

Fig. 7.4. The temperature dependence of the adsorption of DPC on kaolin in water [from Ref. (12)].

References

1. Shinoda, K., and M. Fujihara, *Bull. Chem. Soc. Jpn. 41*:2162 (1968).
2. Shinoda, K., *J. Phys. Chem. 81:*1300 (1977).
3. Shinoda, K., *Principles of Solution and Solubility,* Marcel Dekker, New York, Chpt. 7 (1978).
4. Costas, M., B. Kronberg, and R. Silveston, *J. Chem. Soc. Farady Trans. 90:*1513 (1994).
5. Kronberg, B., M. Costas, and R. Silveston, *J. Disp. Sci. & Technol. 15:*333 (1994).
6. Kronberg, B., M. Costas, and R. Silveston, *Pure Appl. Chem. 67:*897 (1995).
7. Kronberg, B., and R. Silveston, *J. Phys. Chem. 93:*6241 (1989).
8. Patterson, D., and M. Barbe, *J. Phys. Chem. 80*:2345 (1976).
9. Adderson, J.E., and H. Taylor, *J. Colloid Sci. 19:*495 (1964).
10. Kronberg, B., *J. Colloid Interface Sci. 96:*55 (1983).
11. Kronberg, B., and P. Stenius, *J. Colloid Interface Sci. 102:*410 (1984).
12. Mehrian, T., A. de Keizer, and J. Lyklema, *Langmuir 7:*3094 (1991).

Chapter 8

Gemini Surfactants

Milton J. Rosen

Surfactant Research Institute, Brooklyn College, City University of New York

Several years ago, our Surfactant Research Institute at Brooklyn College was working on a project for the Dow Chemical Company in which we were determining the structure/property relationships of the various types of surfactants that were present in their Dowfax surfactants. The types of surfactants that could be present in Dowfax are shown in Fig. 8.1. Surfactants of these four types—well purified and well characterized—were synthesized for us in Dow's research laboratories. In the course of investigating the surface properties of those materials (1), we realized that the properties of the DADS (dialkyl diphenylether disulfonate) type of surfactant were very unusual. This type of surfactant was much more surface active than we had expected. Its tendency to adsorb at an interface and to form micelles was much greater than that of a conventional surfactant of comparable structure (i.e., having a similar [single] hydrophilic headgroup and an equivalent [single] hydrophobic group).

At about the same time, a group of Japanese investigators at Osaka University started to publish data on surfactants having two hydrophilic and two hydrophobic groups in the molecule (2–6), and we noticed that their molecules also were much more surface active than expected. Eventually, we joined forces and together published sev-

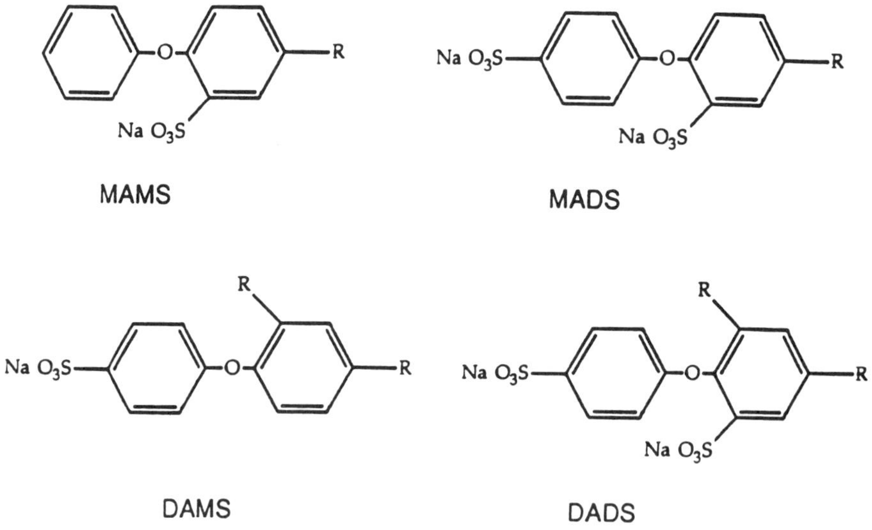

Fig. 8.1. Dowfax components.

eral papers on these surfactants (7–10). A few years earlier, some diquaternary ammonium compounds with two hydrophobic groups in the molecule had been synthesized by some other investigators (11–13) and their critical micelle concentrations (CMC) measured. Although these CMC values were much lower than those of comparable conventional surfactants, this appears not to have been commented upon by these investigators.

We have now investigated several homologous series of geminis (surfactants having two hydrophilic groups and two or three hydrophobic groups), both anionic and cationic (Rosen, M.J., L. Liu, and L.D. Song, (14–16), and have found that they indeed have some very unusual and unexpected properties.

Properties of Gemini Surfactants

Table 8.1 shows some interfacial properties of some dianionic gemini surfactants, together with those of some conventional surfactants having similar (single) hydrophilic groups and roughly equivalent (single) hydrophobic groups in the molecule. It is apparent from the data that the geminis have CMC values 1–2 orders of magnitude smaller and C_{20} values (the molar surfactant concentration in the aqueous phase required to decrease the surface tension of the solvent by 20 dyn/cm, a measure of surfactant efficiency) 2–3 orders of magnitude smaller than the comparable conventional surfactants. Note that the surface tension at the CMC (γ_{CMC}) of the geminis is at least as low as, if not better than, that of the conventional surfactants. Table 8.2 shows data for some diquaternary ammonium cationic geminis, $[C_nH_{2n+1}N(CH_3)_2CH_2CHOH]_2^{2+}$ • 2Br–, synthesized in our laboratory (14,16), together with those of comparable conventional surfactants. Their CMC values are again 1–2 orders of magnitude and their C_{20} values 2–3 orders of magnitude smaller than those of the latter, while their γ_{CMC} values are at least as low. Noteworthy is the effect of the structure of the linkage between the two hydrophilic groups. The CMC values are smallest when the linkage is flexible and hydrophilic ($-CH_2CHOHCHOHCH_2-$), larger when the linkage is flexible and hydrophobic ($-CH_2CH_2CH_2CH_2-$), and somewhat larger when it is rigid and hydrophobic ($-CH_2C_6H_4CH_2-$). Even with a rigid hydrophobic linkage, however, the CMC values are still 1–2 orders of magnitude smaller than those of the comparable conventional surfactants.

We have also found that the packing of the hydrophobic groups in the geminis at the aqueous solution/air interface (i.e., the surface area/chain) is closer than that found in the comparable conventional surfactants, which may account for the observation that the surface tension at the CMC of the solution is often lower in the former than in the latter.

Why are the gemini surfactants so much more surface active than the comparable conventional surfactants? One would normally expect that when the number of alkyl chain carbon atoms per hydrophilic head group is the same in both conventional surfactants and in geminis, i.e., the hydrophile-lipophile balance in both compounds is the same, that their surface properties would be similar. The answer lies in the number of alkyl chain carbon atoms per molecule. It is well known that an increase in the number of carbon atoms in the alkyl chain of a (conventional) surfactant increases its surface

TABLE 8.1 Properties of Some Anionic Gemini-Type Surfactants in Water at 25°C[a]

Type	Y	CMC (mM)	γ_{CMC} (dyn/cm)	C_{20} (mM)
A	$-OCH_2CH_2O-$	0.013	27.0	0.001
$C_{12}H_{25}SO_4Na$	—	8.2	39.5	3.1
B	$-O-$	0.033	28.0	0.008
B	$-OCH_2CH_2O-$	0.032	30.0	0.0065
B	$-O(CH_2CH_2O)_2-$	0.060	36.0	0.001
$C_{12}H_{25}SO_3Na$	—	9.8	39.0	4.4

[a]Source: Ref. 3,4.

TABLE 8.2 Surface Properties of Some Cationic Geminis in Water at 25°C[a]

Compound	CMC (mM)	γ_{CMC} (dyn/cm)	C_{20} (mM)
$(C_{10}NMe_2CH_2CHOH)_2^{2+} \cdot 2Br-$	3.7	35.5	0.58
$(C_{10}NMe_2CH_2CH_2)_2^{2+} \cdot 2\,Br-$	9.0	38.5	—
$C_{10}NMe_3^+Br-$	68	—	—
$(C_{12}NMe_2CH_2CHOH)_2^{2+} \cdot 2Br-$	0.7	35.4	0.13
$(C_{12}NMe_2CH_2CH_2)_2^{2+} \cdot 2Br-$	1.2	—	—
$[(C_{12}NMe_2CH_2)_2C_6H_4]^{2+} \cdot 2Br-$ (50°C)	1.3	39.5	—
$C_{12}NMe_3^+Br-$	16	—	8
$(C_{14}NMe_2CH_2CHOH)_2^{2+} \cdot 2Br-$	0.085	36.0	0.0028
$C_{14}NMe_3^+Br-$	3.6	—	—

[a]Source: Ref. 14,16.

activity, i.e., decreases its CMC and C_{20} values. This is due to the increased distortion of the water structure by the increased length of the alkyl chain, and its distortion of the solvent structure is the basis for the surface activity of the surfactant (17). However, this increase also decreases the solubility of the surfactant in water and this limits the increase in the length of the chain (and hence, the increase in surface activity) of the conventional surfactant in aqueous systems. With two hydrophilic groups, however, the solubility of the gemini in water is increased greatly, and this permits the gemini molecule to contain many more alkyl chain carbon atoms and still remain water soluble, with the resulting great increase in surface activity.

What does this increased surface activity of the geminis mean for their practical utilization?

1. The higher surface activity means that less surfactant may be needed to perform a function for which surfactants are used, e.g., detergency or emulsification. This means less raw material needed for the synthesis, less manufacturing by-product to be handled, and less environmental impact of the smaller surfactant quantity used to perform the particular function.
2. The much lower CMC of the gemini means that it may produce less skin irritation, because irritation is usually a function of the concentration of monomeric surfactant in the solution phase, and this decreases with a decrease in the CMC. A very low CMC value also means that such properties as solubilization of water-insoluble material can occur at much lower surfactant concentrations with geminis than with comparable conventional surfactants, because solubilization occurs only when the CMC has been exceeded.
3. The double charge on an ionic gemini means that it should interact more strongly (attractively) with surfactants of neutral and opposite charge than do conventional surfactants. Because such attractive interaction between surfactants is the basis for synergy in their mixtures, this means that stronger synergy can be expected to be exhibited in mixtures of a gemini than of a comparable conventional surfactant with an oppositely charged or neutral surfactant.

Some surface tension data are shown in Fig. 8.2. Values of the interaction parameters (β parameters) for various systems are shown in Table 8.3. The first noteworthy feature is that the mole fractions of ≈ 0.5 of the two surfactants, both in the mixed monolayer (X^m) at the aqueous solution/air interface and in the mixed micelle (X^m) in the aqueous phase indicate that the two surfactants are interacting in a 1:1 molar ratio, rather than in the 2:1 molar ratio that would be expected of a divalent gemini and a monovalent conventional surfactant. As a result of this 1:1 molar interaction, the complex formed is water soluble, because there is a net charge on it. This is in contrast to what occurs in mixtures of conventional anionic and cationic surfactants in which the 1:1 molar interaction product is a neutral complex that is often water insoluble and precipitates from solution.

The second point of interest in the data is the much weaker interaction between the gemini and the second surfactant in the mixed micelle (β^m), compared with their interaction in their mixed monolayer at the aqueous solution/air interface (β^σ). This is generally observed in gemini-containing mixtures and may be due to the greater difficulty of incorporating the two hydrophobic groups of the geminis into a convex micelle than of accommodating them at a planar interface.

Properties of Mixed Surfactant Systems

Hydrotropic properties. The hydrotropic properties of surfactants stem from their ability to inhibit the formation of crystalline or liquid crystalline structures in the

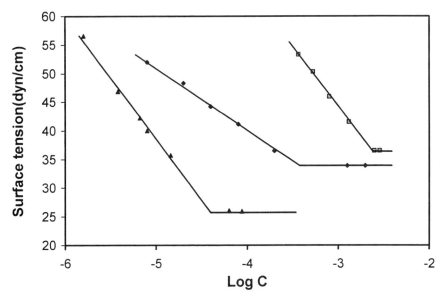

Fig. 8.2. Synergism in the system $(C_{10}NMe_2CH_2CHOH)_2{}^{2+} \cdot 2\ Br^- - C_{12}H_{25}SO_3Na$ in 0.1 M NaBr at 25°C. *Source:* Ref. 15. ■, $C_{12}H_{25}SO_3Na$; ♦, $(C_{10}NMe_2CH_2CHOH)_2{}^{2+} \cdot 2\ Br^-$ (C_{10} Diq); ▲, C_{10} Diq $- C_{12}H_{25}SO_3Na$ ($\alpha = 0.67$).

TABLE 8.3 Interaction Parameters for the Systems $(C_nNMe_2CH_2CHOH)_2{}^{2+} \cdot 2\ Br^- - C_nSO_3{}^-Na^+$ at 25°C[a]

System	Medium (0.1 M)	α	β$^\sigma$	β^m	X	X^m
C_8Diq[b]-$C_{10}SO_3$Na	NaBr	0.40	−26	−12	0.49	0.51
C_8Diq[b]-$C_{12}SO_3$Na	NaBr	0.40	−26	−12	0.49	0.51
C_{10}Diq[b]-$C_{10}SO_3$Na	NaBr	0.40	−26	−12	0.49	0.51
C_{10}Diq[b]-$C_{12}SO_3$Na	NaBr	0.40	−26	−12	0.49	0.51
C_{10}Diq[b]-$C_{12}SO_3$Na	NaBr	0.40	−26	−12	0.49	0.51

[a]*Source:* Ref. 15.
[b]$(C_nNMe_2CH_2CHOH)_2{}^{2+} \cdot 2\ Br^-$.

aqueous phase. Compounds that have a large hydrophilic portion relative to their hydrophobic portion are often good hydrotropes. Table 8.4 shows some data for the four types of alkylated diphenylether sulfonates (C_{10}DPE sulfonates) shown in Fig. 8.1, where R is a linear C_{10} alkyl chain (attached to the aromatic ring at different nonterminal carbon atoms). The data are for the addition of the different diphenylether sulfonates to a pasty, opaque, aqueous dispersion of linear alkylbenzene sulfonate (LAS). In each case, the added diphenylether sulfonate was 10% of the weight of the LAS (on a 100% active basis). As expected, the monoalkylated disulfonate (MADS), which has the largest ratio of hydrophilic:hydrophobic regions, is the best hydrotrope. However, the gemini (DADS), with its two hydrophilic groups, is a better hydrotrope than the (conventional-structured) monoalkylated monosulfonate (MAMS) or the dialkylated monosulfonate (DAMS).

Solubilization. To evaluate the solubilization properties of the C_{10} DADS gemini, relative to those of the other Dowfax components, the absorbance in the visible (at 600 nm) of mixtures of these materials with various water-insoluble surfactants at 1 g/L total surfactant concentration was measured. Data are shown in Table 8.5. In the 1-cm cell in which the absorbances were measured, an absorbance of 0.008 is given by an aqueous solution that is clear to the naked eye with only a trace of translucence; an absorbance of 0.020, by a solution with a slight translucence; and 0.100 or more, by cloudy solutions.

TABLE 8.4 Hydrotropy of C_{10} Diphenylether (DPE) Sulfonates (23% LAS + 2.2% C_{10}DPE sulfonate)[a,b]

C_{10} DPE Sulfonate	Appearance
None	Viscous, cloudy dispersion at 25°C
MADS	Clear, flowable at 15°C
DADS	Clear, viscous at 25°C; cloudy at 15°C
MAMS	Viscous, cloudy dispersion at 25°C
DAMS	Viscous, cloudy dispersion at 25°C

[a]*Source:* Ref. 1.
[b]*Abbreviations:* MADS, monoalkylated disulfonate; DADS, dialkylated diphenylether disulfonate; MAMS, monoalkylated monosulfonate; DAMS, dialkylated monosulfonate.

TABLE 8.5 Absorbance at 600 nm of C_{10} Diphenylether (DPE) Sulfonate Solutions of H_2O-Insoluble Surfactants (conc., 1 g/L)[a]

	H_2O-Insoluble Surfactant [b]		
C_{10}DPE Sulfonate (conc., g/L)	Sil	AE	APE
DADS (0.1)	0.008	0.008	0.006
MAMS (0.1)	0.008	0.007	0.008
MADS (0.1)	0.025	—	—
DADS (0.1)	0.020	0.008	0.014
MAMS (0.01)	0.011	0.023	0.079
DAMS (0.01)	0.070	0.044	0.145

[a]*Source:* Ref. 1.
[b]*Abbreviations:* Sil, siloxane-type surfactant; AE, alcohol ethoxylate; APE, alkylphenol ethoxylate. Other abbreviations as in Table 8.4.

It can be seen from the data that the DADS gemini, especially at 1% addition to the insoluble surfactant, is a better solubilizer than the other Dowfax components. This has been confirmed in our laboratory with other geminis, compared with comparable conventional surfactants, and is probably due to the very low CMC values of the geminis because, as mentioned above, solubilization commences only after the CMC has been exceeded.

Enhancement of wetting. The solubilization of certain water-insoluble surfactants in aqueous medium often increases their wetting properties considerably (18). Table 8.6 lists some data on this phenomenon. The data are all for 1 g/L total surfactant concentration in water at 25°C, with Draves skein wetting times (WOT) in seconds. Absorbances of the solutions at 600 nm (in a 1-cm cell), equilibrium surface tension (γ_{eq}) in dyn/cm and dynamic surface tension at 1-s surface age (γ_{1s}) in dyn/cm, are also listed. It is noteworthy that there is no correlation between the equilibrium surface tension value (γeq) and the wetting time (WOT). Even when γ_{eq} is very low (26–27 dyn/cm), WOT can be very high if the solution is cloudy. The main correlation is between the surface tension at 1-s surface age (γ_{1s}) and WOT (18).

In the $C_{12}EO_3$ (*n*-dodecylmonoether of trioxyethyleneglycol)-$C_{12}EO_8$ (*n*-dodecylmonoether of octaoxyethyleneglycol) system, the replacement of increasingly greater amounts of the water-insoluble $C_{12}EO_3$ by the water-soluble $C_{12}EO_8$ makes the aqueous phase somewhat clearer and decreases the wetting time considerably. However, all WOT values fall between those of the two individual surfactants by themselves, indicating no synergy. The addition of a conventional sulfonate surfactant, $C_{12}H_{25}SO_3Na(C_{12}SO_3Na)$, produces somewhat similar effects, increasing the clarity of the solution somewhat and decreasing the WOT value of the $C_{12}EO_3$ considerably, but here the known stronger interaction of $C_{12}SO_3Na$ than $C_{12}EO_8$ with $C_{12}EO_3$ results in synergism in dynamic surface tension reduction (γ_{1s} for the mixtures is smaller than the γ_{1s} values for the individual surfactants by themselves) and this produces synergism in Draves skein wetting (WOT values for the mixtures are smaller than for the individual surfactants by themselves).

The addition of the gemini disulfonate ($C_{10}DADS$), which because of its double negative charge is expected to interact more strongly with $C_{12}EO_3$ than $C_{12}SO_3Na$, produces stronger synergy in dynamic surface tension (γ_{1s}) and in wetting (WOT) than the addition of the latter. Although the $C_{12}EO_3$ by itself has a WOT of 129 s and the $C_{10}DADS$ by itself has a WOT of 431 s, replacement of 20% of the $C_{12}EO_3$ by $C_{10}DADS$ yields a WOT of only 14.5 s.

A similar dramatic decrease in wetting time is observed when $C_{10}DADS$ is added to the commercial nonylphenol ethoxylate (Igepal CO-430), which is insoluble in water and has a WOT of 114 s. Replacement of 20% of Igepal by $C_{10}DADS$ results in a wetting time of 11 s, even though the $C_{10}DADS$ by itself is a very poor wetting agent.

An even more dramatic effect is observed upon the replacement of 20% of *n*-dodecyl pyrrolidone ($C_{12}P$), a water-insoluble nonionic surfactant, by $C_{10}DADS$.

TABLE 8.6 Enhancement of Wetting of H_2O-Insoluble Surfactants (total surfactant, 1 g/L at 25°C)[a,b]

System	Surfactant ratio (wt/wt)			
	1:0	0.8/0.2	0.5/0.5	0:1
$C_{12}EO_3$-$C_{12}EO_8$				
Abs (600 nm)	cloudy	0.082	0.045	clear
γ_{eq} (dyn/cm)	27.1	27.7	28.1	34.8
γ_{1s} (dyn/cm)	45.9	43.9	35.9	37.3
WOT (s)	129	24.6	14.2	9
$C_{12}EO_3$-$C_{12}SO_3Na$				
Abs (600 nm)	cloudy	0.047	0.047	clear
γ_{eq} (dyn/cm)	27.1	27.0	26.9	54.5
γ_{1s} (dyn/cm)	45.0	37.9	37.7	56.5
WOT (s)	129	19.3	16.8	28.0
$C_{12}EO_3$-$C_{10}DADS$				
Abs (600 nm)	cloudy	0.031	0.005	clear
γ_{eq} (dyn/cm)	27.1	27.0	28.5	44.3
γ_{1s} (dyn/cm)	45.0	35.0	39.1	66.8
WOT (s)	129	14.5	17.5	431
Igepal CO-430-$C_{10}DADS$				
Abs (600 nm)	cloudy	0.024	0.006	clear
WOT (s)	114	11.0	10.0	431
$C_{12}P$-$C_{10}DADS$				
Abs (600 nm)	0.465	0.007	0.006	clear
γ_{eq} (dyn/cm)	26.6	26.8	29.7	44.3
γ_{1s} (dyn/cm)	68.6	33.1	39.0	66.8
WOT (s)	130	8.6	13.2	431

[a]Source: Ref 1.
[b]Abbreviations: Abs, absorbance; WOT, wetting time. Other abbreviations as in Table 8.4.

n-Alkyl pyrrolidones have been shown (19) to be capable of accepting a proton and interacting as a cationic surfactant in the presence of an anionic surfactant. This interaction is stronger than the interaction of $C_{10}DADS$ with a nonionic surfactant incapable of accepting a proton. As a result of this stronger interaction, the replacement of 20% of the $C_{12}P$ by $C_{10}DADS$ produces a clear solution that has a WOT of 8–9 s.

Conclusions Regarding Commercial Utilization of Geminis

From the above data, there appear to be two modes for the commercial utilization of geminis. First, if the gemini can be produced at a price roughly equivalent to that of a commercially utilized surfactant, then the gemini may have a cost effectiveness advantage, because less of the gemini may be needed to perform a particular function due to its higher surface activity. An additional fringe benefit in some cases may be the lower skin irritation of the gemini. Second, if the gemini can be produced only at a much higher cost than a commercially utilized surfactant, then the gemini may be used as a low-percentage additive to the conventional surfactant to enhance its properties, thus justifying the added cost.

Acknowledgments

This presentation is based on work published with my collaborators, Dr. Xi Yuan Hua, Dr. Tao Gao, Zhen Huo Zhu, Letian Liu, and Li Dong Song, and was supported by grants from the Colgate Palmolive Co., Dow Chemical Co., Reckitt and Colman, Rhone-Poulenc Surfactants and Specialities, Texaco Inc., Witco Corp., and the National Science Foundation.

References

1. Rosen, M.J., Z.H. Zhu, and X.Y. Hua, *J. Am. Oil Chem. Soc. 69:*30 (1992).
2. Okahara, M., A. Masuyama, Y. Sumida, and Y.-P. Zhu, *J. Jpn. Oil Chem. Soc. (Yukagaku) 37:*746 (1988).
3. Zhu, Y.-P., A. Masuyama, and M. Okahara, *J. Am. Oil Chem. Soc. 67:*459 (1990).
4. Zhu, Y.-P., A. Masuyama, T. Nagata, and M. Okahara, *J. Jpn. Oil Chem. Soc. (Yukagaku) 40:*473 (1991).
5. Zhu, Y.-P., A. Masuyama, and M. Okahara, *J. Am. Oil Chem. Soc. 68:*268 (1991).
6. Zhu, Y.-P., A. Masuyama, Y.-I. Kirito, and M. Okahara, *J. Am. Oil Chem. Soc. 68:*539 (1991).
7. Masuyama, A., T. Hirono, Y.-P. Zhu, M. Okahara, and M.J. Rosen, *J. Jpn. Oil Chem. Soc. 41:*301 (1992).
8. Zhu, Y.-P., A. Masuyama, Y.-L. Kirito, M. Okahara, and M.J. Rosen, *J. Am. Oil Chem. Soc. 69:*626 (1992).
9. Zhu, Y.-P., A. Masuyama, Y. Kobata, Y. Nakatsuji, M. Okahara, and M.J. Rosen, *J. Colloid Interface Sci. 158:*40 (1993).
10. Rosen, M.J., T. Gao, Y. Nakatsuji, and A. Masuyama, *Colloid and Surface A 88:*1 (1994).
11. Parreira, H.C., E.R. Lukenbach, and M.K.O. Lindemann, *J. Am. Oil Chem. Soc. 56:*1015 (1979).
12. Devinsky, F., L. Masarova, and I. Lacko, *J. Colloid Interface Sci. 105:*235 (1985).
13. Devinsky, F., I. Lacko, F. Bittererova, and L. Tomeckova, *J. Colloid Interface Sci. 114:*314 (1986).
14. Rosen, M.J. and L. Liu, *J. Am. Oil Chem. Soc. 73*:885 (1996).
15. Liu, L. and M.J. Rosen, *J. Colloid InterFace Sci. 179*:454 (1996).
16. Song, L.D. and M.J. Rosen, *Langmuir 12*:1149 (1996).
17. Rosen, M.J., *Surfactants and Interfacial Phenomena,* 2nd edn., Wiley, New York, 1989, p. 3.
18. Rosen, M.J. and Z.H. Zhu, *J. Am. Oil Chem. Soc. 70:*65 (1993).
19. Zhu, Z.H., D. Yang, and M.J. Rosen, *J. Am. Oil Chem. Soc. 66:*998 (1989).

Chapter 9

Surfactants in the Environment

John F. Scamehorn, Sherril D. Christian, Jeffrey H. Harwell, and David A. Sabatini

Institute for Applied Surfactant Research, The University of Oklahoma, Norman, OK 73071

Introduction

Surfactants are used in an increasing number of applications in protecting or cleaning up the environment. In this paper, a method to remove pollutants from wastewater or groundwater, as well as the use of surfactants in remediating contaminated soils, will be discussed as two examples of applications of surfactants in the rapidly growing environmental field.

Organic contaminants or inorganic ions can be removed from aqueous streams by micellar-enhanced ultrafiltration (MEUF). In MEUF, surfactant is added to the water at concentrations well above the critical micelle concentration (CMC). Organic pollutants solubilize in the micelles while multivalent ions of opposite charge to that on the surfactant bind to the micelle surface. The solution is then treated by ultrafiltration with membranes having pore diameters small enough to block the passage of micelles, resulting in purified water and a concentrated waste stream which can be treated to recover the surfactant for reuse. The process can be staged to attain any degree of purification. Pollutants which have been removed from water by this technique include phenolics, chlorinated hydrocarbons, heavy metals, and chromate.

The use of surfactants to remediate contaminated soil has been shown to be useful in removing dense, non-aqueous phase liquids (DNAPLs) in laboratory tests, and large field tests are imminent. Two different basic mechanisms can be used to remove these pollutants: solubilization in micelles and mobilization of the oil by attaining ultralow interfacial tensions. It is not clear which of these mechanisms will end up being the most effective in commercial operations. While much effort has been expended to find edible surfactants which can be used, this class has been shown to be less effective than a synthetic anionic surfactant in removal of trichloroethylene. The surfactant must have high solubilization capacity, high hardness tolerance, and low adsorption onto soils.

Micellar-Enhanced Ultrafiltration

Ultrafiltration (UF) is a useful separation process for removing and recovering solute species having molecular weights of 1,000 Daltons or more. Because high fluxes can be obtained at relatively low pressures, UF is particularly attractive as a low-energy industrial method for separating relatively large molecules from water. Unfortunately, traditional UF is not effective in removing solutes having molecular weights less than about 500 Daltons. In micellar-enhanced ultrafiltration, pollutant molecules in water which are too small to be removed by direct ultrafiltration are attached to these micelles and the micelles ultrafiltered from solution, resulting in purified water. The economic advantages of MEUF are that it is a low-energy process, it may be staged to achieve any desired degree of separation, it can simultaneously remove both nonionic organic molecules and charged ions (inorganic or organic), and it uses a nontoxic, biodegradable separating agent (surfactant), resulting in quite small residual concentrations of the separating agent in the product water. MEUF can be applied in treating water from such diverse sources as: chemical process industries (water containing various dissolved organic components such as chlorinated hydrocarbons, aromatics, and other highly toxic species), metal plating industries (metal ions such as zinc, nickel, and chromate), synfuels plants (organics such as phenol derivatives, and metals, such as lead), refineries (both organics and metal ions), farmland runoff (insecticides, fertilizers, and multivalent metal ions), metal and coal mine leachates (heavy metal ions and organics), nuclear reactors (ionic fission products), and industrial laundries and hard-surface cleaning (benzene, chlorinated hydrocarbons, and metal ions).

Dissolved organic solutes (1–14) or multivalent ions (6,7,9–11,15–24) can be removed from water using MEUF. This is illustrated in Fig. 9.1 for removal of multivalent cations and organic solutes. In this example, an anionic surfactant is first added to the water at concentrations well above the CMC (i.e., the concentration at which aggregates of surfactant molecules called micelles begin to form in significant amounts). For many surfactants in aqueous solution, the CMC is in the range 0.00001–0.01 M, and the micelles contain 50–150 molecules of surfactant.

The interior of a micelle is hydrocarbon-like and capable of "solubilizing" organic solutes. When anionic surfactants are used in MEUF, the head group regions of the micelle are negative, and the micelle surface has a highly negative electrical potential. Therefore, multivalent metal ions in solution (e.g., copper, lead, or thorium) preferentially bind to the oppositely charged micellar surface by electrostatic attraction. Conversely, when cationic surfactants are used, negative ions such as chromate will bind to the positively charged micellar surface. If a feed solution containing the added surfactant is forced through an ultrafiltration membrane having pore diameters small enough to block the passage of the micelles, the organic molecules and metal ions bound to the micelles are also prevented from entering the stream that passes through the membrane (the permeate). The concentrations of the organic molecules, metal ions, and surfactants in the permeate correspond very nearly to their uncomplexed species concentrations in the solution retained by the membrane (the retentate) (1,2,5,6,8,16).

In a properly designed MEUF process, the permeate concentrations of all of the target ions will be quite small; the solute in the permeate can often have a concentration two orders of magnitude smaller than in the feed (8,17). In many cases, the permeate stream can in many cases be discharged directly into the environment or reused in an industrial plant. The concentrated waste solution in the retentate will contain almost all of the original solutes in a volume which can be a small fraction of the orig-

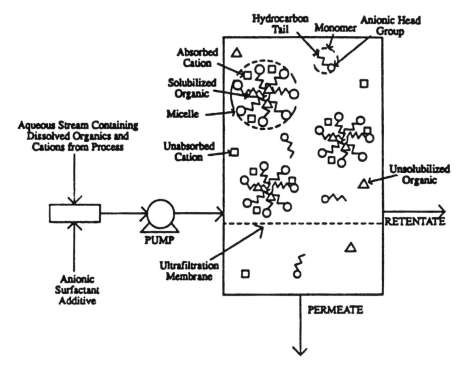

Fig. 9.1. Schematic diagram of micellar-enhanced ultrafiltration (MEUF) for organic solute and cationic metals.

inal feed water. MEUF can be staged to attain any required degree of separation (6,14).

Table 9.1 includes results obtained by using a one-stage MEUF process to remove both organic solutes and inorganic ions: anionic surfactants are used to remove cations; cationic surfactants are used to remove anions; and either class of surfactant can be employed to remove neutral species. For almost all of the solute species, low permeate concentrations and quite large rejections are observed. The organic solutes of low solubility tend to be removed most efficiently in MEUF; e.g., compounds in a homologous series containing longer alkyl chains will exhibit higher rejections.

A major result of our studies of the use of MEUF in removing organic solutes is that the concentration of a solute passing from the retentate into the permeate is practically equal to the concentration of the monomeric or free solute, i.e., the concentration of the solute not bound to micelles. Therefore, one can use information about the equilibrium solubilization of solutes by surfactant micelles to predict *a priori* the efficiency of MEUF in removing these solutes from aqueous streams (1,2,4–8). A recent book summarizes the current state of knowledge of the phenomenon of solubilization (25), and a review article summarizes methods of measuring solubilization (26). An auto-

TABLE 9.1 Comparison of Permeate Concentrations of Organic and Metal Pollutants in MEUF[a]

Pollutant	Pollutant concentration in permeate (mM)	Rejection (%)
Phenol	1.42	94.3
m-Cresol	0.526	97.9
4-tert-Butylphenol	0.0767	99.7
Benzene	2.33	90.7
Toluene	0.80	96.8
n-Heptanol	0.323	98.7
n-Octanol	0.141	99.4
Zinc (+2)	0.037	99.9
Copper (+2)	0.037	99.9
Cadmium (+2)	0.037	99.9
Chromate (–2)	0.030	99.9

aConditions: Retentate [pollutant] = 25 mM; retentate [surfactant] = 250 mM; pressure = 414 kPa; membrane = 1 K to 20 K molecular weight cutoff; temperature = 30°C; rejection (%)= 100{1-[permeate]/[retentate]}.

mated vapor pressure apparatus (4,27–33) or traditional head-space chromatography (34) can be used to generate solubilization isotherms (equilibrium solubilization levels as a function of micellar composition) for volatile solutes. A more general method, semiequilibrium dialysis (SED) (5,8,35–43) can measure solubilization isotherms for volatile or nonvolatile solutes. Several hundred solubilization isotherms have been obtained for various organic solutes, including many substances that are dangerous pollutants in wastewaters (44).

Mass action models have been used to correlate and explain solubilization results for systems for which extensive results of sufficient accuracy are available (32,33). The pseudo-phase equilibrium theory has also been used to describe complicated aqueous solutions containing surfactants and organic solutes and to understand the behavior of solubilization isotherms at various temperatures and various concentrations of surfactants (1,2,4,5,8,27–32,35–41). Finally, a group contribution method has been used to predict the solubilization constants for organic solutes of arbitrary structure in anionic or cationic surfactant micelles (8,45). As a result, procedures are now available for estimating the efficiency of removal of organic solutes in actual MEUF separations, using only knowledge of the structural formulas of these compounds.

MEUF can be used to remove either anionic or cationic multivalent ionic species that bind to surfactant micelles (6,7,9,10,11,15–24). For example, we have investigated the removal of a number of cations, including zinc, nickel, copper, lead, and calcium, and several anions, including chromate, and the cyano-complexes of iron and nickel. A simple equilibrium model, combining the two-phase polyelectrolyte theory of Oosawa (46) with thermodynamic activity, material-balance, and charge-balance equations has been quite successful in predicting the ability of MEUF to remove dissolved cations and anions from aqueous streams (16,17,11,20,24). The equilibrium assumption mentioned above, which has been valuable in predicting the efficiency of MEUF in removing dissolved organic species, seems to apply equally well in estimating the concentrations of ionic species passing through the ultrafiltration membrane. Results from SED experiments produce results in essential agreement with MEUF data for several aqueous ionic solute/ionic surfactant systems (15–17). Organic solutes and metal ions can be removed simultaneously in MEUF processes using anionic surfactants without a loss of efficiency compared with removal of only one of the contaminants alone (10). The ability to replace two unit operations by a single unit operation to remove the two dissimilar pollutants can make MEUF particularly attractive economically.

Concentration polarization effects have been studied in detail for stirred cell and spiral wound ultrafiltration units, and gel points have been determined for a number of surfactant types (1,2,17,14,24). In Fig. 9.2, relative flux data are plotted as functions of retentate surfactant concentration, for an anionic surfactant (sodium dodecylsulfate), a cationic surfactant (cetylpyridinium chloride), and for comparison, similar data for another colloid, a polyelectrolyte (sodium polystyrenesulfonate) (47). These results show that the relative fluxes remain large until the retentate solution becomes quite concentrated (>0.3 M surfactant), that the concentration polarization obeys accepted gel polarization ultrafiltration models [i.e., flux vs. log(retentate concentration) is linear], and that gel point concentrations are quite large (*ca.* 0.6 M colloid). Electrostatic repulsion between micelles or polyelectrolytes may help prevent coagulation in the gel layer next to the membrane, so that gel polarization is not severe in these systems. This observation is supported by the fact that poorer fluxes are obtained when using nonionic surfactants in MEUF separations (14). It has been shown that the rejection of either organic solutes or multivalent heavy metals remains high as the surfactant concentration in the retentate increases as long as these concentrations are below about 0.3 M for the surfactants studied (2,24). Because flux levels are unacceptably low above this concentration, the reduction in rejection (believed to be due to the formation of premicellar aggregates) at these retentate compositions is not a limitation.

Leakage of surfactant monomer molecules into the permeate can cause problems due to the cost of replacing the lost surfactant or legal restrictions on allowable effluent surfactant concentrations. The use of polyelectrolyte/surfactant complexes exploits the fact that the concentration of monomeric surfactant in equilibrium with these colloidal aggregates is greatly reduced (compared with surfactant micelles) so that the subsequent leakage of surfactant into the permeate is also greatly decreased (37,48). An interesting class of soluble polyelectrolytes, the polysoaps, has both hydrophobic characteristics and ion-binding properties similar to aqueous surfactant micelles; thus, they may yield excellent separation efficiencies without significant leakage into the permeate.

Another approach is to remove the surfactant which leaks into the permeate in a downstream operation. If a sufficiently high molecular weight surfactant is used, ultrafiltration using small pore size membranes can be used to remove this residual surfac-

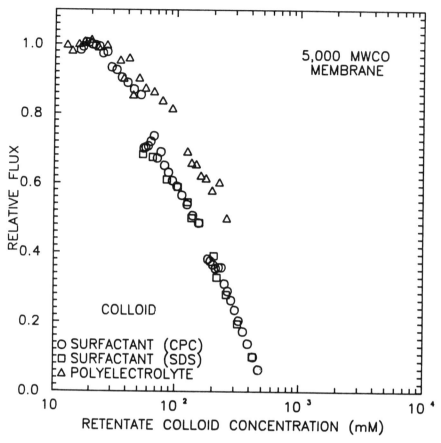

Fig. 9.2. Relative fluxes for a cationic surfactant, an anionic surfactant, and an anionic polyelectrolyte in ultrafiltration.

tant from the permeate (14). Foam fractionation is a technique which is highly effective in removing the surfactant from the permeate (49). Oppositely charged polyelectrolytes can be used to complex ionic surfactant molecules which escape into the permeate, and this complex can be removed subsequently by ultrafiltration.

Economic analysis of MEUF shows that a high fraction of the surfactant in the retentate must be recovered for reuse for economical use of this technology in most applications. The method used to recover the surfactant is highly dependent on the solutes being removed from the water, as well as the type of surfactant used (50). If the pollutant is volatile, stripping (vacuum, steam, or air) can remove the solute from the retentate, allowing the contaminant-free, concentrated surfactant solution to be reused. The stripper overheads can be condensed, incinerated, or treated in some other fashion. If an ionic surfactant is used, it can be precipitated from solution by addition of a counterion or by a reduction in temperature. After removal by settling, filtration, or centrifugation, the surfactant can be redissolved and reused. The contaminant-laden solution can be disposed of or the pollutant recovered if it is economically justified. Solvent extraction of either the pollutant or the surfactant from the retentate is another method of separating pollutant from surfactant.

One specific example of an application of MEUF which has been analyzed in detail will be mentioned to illustrate use of this technology (14). This application involves removal of 99% of trichloroethylene (TCE) from polluted groundwater where it is present at 1–10 ppm. In this case, four MEUF stages were found to be nearly optimum. The retentate solutions are vacuum stripped to remove the TCE as overhead condensate, leaving a concentrated surfactant solution for recycle. A final polishing step is required to remove residual surfactant from the final permeate. This is accomplished by using a low molecular weight cut-off membrane in a final UF step, producing high-quality water and simultaneously recovering surfactant for reuse. The more stages used, the smaller the total retentate stream size from all of the MEUF stages to be vacuum stripped, but the higher the capital cost. As the number of stages increases at a constant total retentate stream size to be vacuum stripped, the concentration of TCE in the effluent water decreases. This illustrates the process engineering optimization which can be done in the design of a MEUF process for a specific application.

Use of Surfactants in Soil Remediation

Surfactants are generating an enormous amount of interest in the area of soil and ground water remediation (51). This interest derives from the realization in the environmental community that the remediation technologies of the last three decades have not demonstrated an ability to clean the contaminated parts of our environment. Such a realization has produced a call to investigate innovative remediation technologies. Surfactants are particularly attractive for enhancing soil and ground water remediation rates because many types are known to be nontoxic, biodegradable, and effective at low concentrations.

With the door opened to the use of surfactants in remediation, many applications have been proposed, ranging from using foams to using layers of adsorbed cationic surfactants to reduce the mobility of aquifer contaminants (52). By far the greatest interest, however, focuses on the use of surfactants to enhance the efficiency of traditional pump-and-treat technology. Most of the interest to date has been on a process significantly different from that used in the enhanced oil recovery (EOR) technology developed during the 1970s and 1980s, although some researchers believe that some kind of an EOR-like technology will prove most effective and most economical. The non-EOR-like technology being considered consists of circulating micelles through a contaminated zone in an aquifer without producing a sufficiently low interfacial tension to mobilize residual liquid drops of the contaminant; the EOR-like technology consists of producing a sufficiently low interfacial tension to allow the contaminant to flow in the form of droplets toward an extraction well.

If surfactant-enhanced pump-and-treat technology becomes widely used for remediation projects, it will probably prove to be applicable to nearly as wide a range of contaminant types as MEUF; current work already includes applications for fuels, heavy metals, and chlorinated solvents as the contaminants. When the contaminant in an aquifer is a fuel or a heavy metal, however, promising competing technologies exist. It is the case of contamination by low volatility, high density solvents that is attracting the most attention both because this is a widespread problem and because no competing technology shows much promise at present. These types of contaminants are referred to in the environmental literature as DNAPLs, Dense Non-Aqueous Phase Liquids. Common examples found in aquifers include perchloroethylene (PCE), trichloroethylene (TCE), and other chlorinated solvents with specific gravities >1.

DNAPLs are a particularly pernicious form of ground water contamination because they are liquids, and because they are more dense than water, they may penetrate deeply into aquifers; these properties, combined with the low water solubilities and high toxicities of many DNAPLs, make them a major remediation problem (53). When a surface or near-surface release of a Non-Aqueous Phase Liquid (NAPL) contaminates ground water, the contamination occurs in three locations: in the vadose or unsaturated zone (the soil from just below the surface down to where the capillaries in the soil are saturated with water), in or just below the capillary fringe (where the capillaries in the soil are water saturated), or in the aquifer itself, below the water table. If the liquid is a Light Non-Aqueous Phase Liquid (LNAPL) such as gasoline or jet fuel, the non-aqueous liquid phase stays at the top of the aquifer, though the seasonal rise and fall of the water table may result in a residual of non-aqueous liquid below the water table at times (Fig. 9.3). Nevertheless, large volumes of ground water are contaminated only by those components which can dissolve into the ground water itself and be carried away from the site of release by the flow of the ground water—water in aquifers flows down hill just as does water in rivers or streams, only more slowly. The contaminated water that moves away from the site of the spill constitutes the contaminant plume; still, for an LNAPL spill, the zone of residual saturation of the contaminant liquid is relatively small and confined to the upper region of the aquifer. When the contaminant is a DNAPL, however, the liquid may not stay at the top of the aquifer. Figure 9.4 illustrates the situation in which the volume of DNAPL spilled exceeds the volume of the liquid that can be retained by the vadose zone: The DNAPL, which is denser than water, penetrates deeply into the aquifer below the water table. If the volume of the spill is sufficient, a region in which the aquifer medium is saturated with the DNAPL may occur at the aquitard, the impermeable layer at the bottom of the aquifer that causes the ground water to be retained. Contamination that occurs in the vadose zone or at the water table can be addressed by excavation, by pumping down to a residual saturation, or by soil venting. A residual saturation by a contaminant liquid below the water table, however, is a long-term source of ground water contamination that currently is unaddressed by any accepted technology and can pollute millions of gallons of ground water.

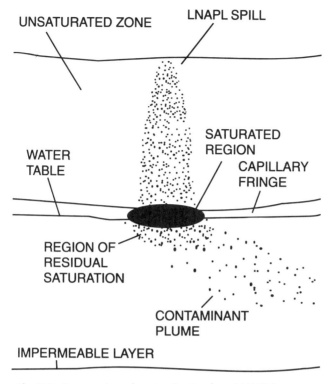

Fig. 9.3. Propagation of contamination by a LNAPL in a ground water environment.

Almost everyone familiar with EOR research immediately assumes that the same technology will be applicable for surfactant-enhanced aquifer remediation (SEAR). This is not the case, however. First, the recovery requirements are different: in EOR, any oil company would be thrilled with a 90% oil recovery rate. In a remediation effort, essentially complete removal of all residual will be necessary in most cases; even a small level of residual liquid can continue to slowly dissolve into the ground water flow, allowing the plume of contamination to continue to grow. For this reason, the small volumes of surfactant solution (a fraction of the volume of the water in the oil reservoir) used in EOR will probably not be used in aquifer remediation. Rather, volumes of surfactant equal to many times the pore volume of the aquifer will most likely be used. The use of fractional pore volumes of solution is necessary in EOR because of the large volumes of oil reservoirs and the overall economics of making money from an EOR project. Because the economics of SEAR restrict one to injecting surfactant only into the zone of residual saturation, this zone will be much smaller, usually on the order of one-half to one-quarter acre; thus, the injection of multiple pore volumes of surfactant is economically feasible, especially if the surfactant is recovered as part of the treatment of the extracted, contaminated water. One restriction

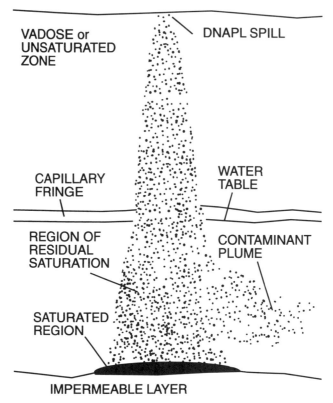

Fig. 9.4. Propagation of contamination by a DNAPL in a ground water environment.

on surfactants used in EOR is therefore removed from SEAR: one need not be overly concerned about chromatographic separation of a mixture of surfactants if optimum behavior can be obtained by blending dissimilar surfactant types; in EOR, chromatographic separation of components could deactivate a small pore volume slug (54,55).

The principal difference between SEAR and EOR, however, is that in EOR the goal was always to mobilize the residual oil saturation by using surfactant to produce an ultralow oil/water interfacial tension. In SEAR, there are definitely situations in which one will not want to mobilize the residual DNAPL. In fact, the majority of the SEAR research performed and published to date has focused on the use of solubilization alone for enhancing the effectiveness of the pump-and-treat operation. In EOR, the density difference between the residual oil and the reservoir brine is generally not sufficient to make buoyancy forces important to the overall process (56). In SEAR, however, if a DNAPL is mobilized and has a significant negative vertical velocity, it could conceivably drop out of the capture zone of the extraction wells, resulting in a spread of the zone of contaminant residual saturation rather than cleanup of the aquifer (57). In SEAR, mobilization will be used only when the aquifer hydrology is favorable to maintaining hydraulic control of the mobilized DNAPL.

When mobilization is not appropriate, micellar solubilization is used to enhance the remediation. This is precisely the type of solubilization used in MEUF for separation of organics from an aqueous process stream. The goal of SEAR in solubilization mode is to circulate as many micelles as possible through the zone contaminated with a residual saturation of the DNAPL. Each micelle present in the circulated solution increases the carrying capacity of the solution being used in the pump-and-treat process. It is possible to design micellar systems that do not produce ultralow interfacial tensions with the residual DNAPL, although the capacity of the micelle for solubilizate decreases as the interfacial tension between the DNAPL and the aqueous phase increases (58).

Many factors have to be considered when selecting a surfactant or a surfactant system for SEAR (59). The surfactant must remain in solution at aquifer conditions. A principal mechanism by which surfactant may be lost from solution in both EOR and SEAR is adsorption. It is easy to underestimate the amount of surfactant that may be lost from solution by this mechanism. Surfactants tend to have high adsorption on most solid surfaces for the same reason that they tend to have higher than anticipated solubilities in solution—they tend to aggregate. Surfactant aggregates form readily on many surfaces (60–62). The cost of the surfactant required to fill the zone of contaminant residual saturation with surfactant at a concentration many times the CMC of the surfactant dominates the economics of SEAR (63). At the adsorption levels of many surfactants, both anionic and nonionic, the amount of surfactant adsorbed may equal or exceed the amount required to flood the contaminated zone with a sufficient concentration of micelles to result in a significantly enhanced remediation rate. For aquifers in which sand or sandstone is the predominant mineral surface present, the surfaces in the aquifer will carry a negative charge; as a result, anionic surfactants will tend to adsorb less under such conditions, especially those with multiple anionic headgroups or with a polyethyleneoxide chain between the tail and the anionic group (64,65). A perusal of Table 9.2 (66) reveals that there are order of magnitude differences in both total adsorptive losses and in the linear distribution coefficient for the dilute surfactant concentration region. The anionics show less adsorption on the soil than the nonionics because the anionic headgroup is repelled by the negative charge on the soil surface which exists near neutral pH. The addition of (EO)-groups to the dodecyl sulfate molecules reduces the adsorption both because the water solubility of the molecule increases and because the headgroup becomes more bulky, occupying more space at the solid/solution interface. When the solid surface is positively charged, as may be the case on limestones or other carbonates, significant reductions in adsorption can be seen when cationic surfactants are used (67).

While anionic surfactant may adsorb significantly less than nonionics on most soils, anionics are susceptible to precipitation whereas nonionics are not. Surfactants with a single, straight alkyl chain as the hydrophobe and with a compact anionic headgroup are especially susceptible to precipitation with multivalent cations. In the

TABLE 9.2 Surfactant Losses from Surfactant Adsorption on an Alluvial Soil[a]

Surfactant	Maximum adsorption (g surfactant/g soil)	Low concentration isotherm slope (cm^3/g soil)
Sodium dodecyl sulfate (SDS)	0.00807	ND[b]
Sodium dodecyl (EO)$_1$ sulfate	0.00097	8.3
Sodium dodecyl (EO)$_2$ sulfate	0.00098	6.3
Sodium dodecyl (EO)$_3$ sulfate	0.00080	4.6
Sodium dodecylbenzene sulfonate	0.0114	8.3
Di(dodecyl)di(phenyl)oxide disulfonate	0.0016	3.1
(EO)$_7$-octylphenol	not evaluated	10.8
(EO)$_8$-nonylphenol	not evaluated	41.1
(EO)$_9$-nonylphenol	not evaluated	31.2
(EO)$_{10}$-nonylphenol	not evaluated	19.7

[a]*Conditions:* Soil-to-water ratio = 1.5 g/mL); background electrolyte, 0.005 M CaCl2; T ca. 22°C; SDS was tested without Ca to avoid precipitation losses.
[b]ND, not determined.

adsorption data shown in Table 9.2, the value for sodium dodecyl sulfate (SDS) is recorded in the absence of calcium because the SDS was observed to precipitate in the presence of low concentrations of calcium at the temperature of the experiments. An important parameter for ionic surfactants is the Krafft point of the surfactant. The Krafft point of a surfactant is the temperature at which the solubility of the monomer reaches the surfactant's CMC; this corresponds to the temperature at which there is a rapid rise in the solubility of the surfactant (68). As a rule of thumb, surfactants are not considered to be water soluble at temperatures below their Krafft point. A surfactant should never be used in a remediation application at a temperature below or near its Krafft point. In such an application, the surfactant would be precipitated from the solution and would fail to make micelles. This means that the lowest aquifer temperature which will be encountered during the remediation project becomes a critical parameter for selecting the surfactant. It must also be remembered that a surfactant's Krafft point will tend to be reduced by multivalent ions. The disulfonates and the ethoxylated sulfonates are highly resistant to precipitation by calcium; some remain in solution at calcium concentrations up to the solubility limit of CaCl$_2$.

Because surfactants are being used to remediate the environment in this technology, it is critical that the surfactants themselves not be pollutants. This requires that they be nontoxic and biodegradable. It is also desirable that they not interfere with the normal biodegradation of the contaminants being removed. Fortunately, most commercial surfactants are highly biodegradable, at least under aerobic conditions. Much less is known through the open literature about anaerobic biodegradation of surfactants. There is an extensive literature on the effects of surfactants on biodegradation; however, the phenomenon is quite complex. It is impossible at present to predict whether a nontoxic surfactant will enhance or inhibit the biodegradation of a particular molecule in a given microbial environment (65,69). The question of surfactant toxicity, however, is much more open to interpretation. While essentially all commercial surfactants are innocuous, not all have undergone the extensive testing required for them to be classified as food and drug additives by the FDA. Gaining regulatory acceptance of a SEAR project in today's complex regulatory environment is facilitated when it can be demonstrated that a given surfactant has either direct or indirect food additive status, even though the purpose of the testing was not to determine the surfactant's toxicity in a SEAR application. As a result, much attention has been given to the use of surfactants with direct food additive status. In general, it can be concluded that such surfactants can be found which will remain active in a ground water environment, will produce acceptable levels of solubilization and adsorption, and can even produce ultralow interfacial tensions between DNAPLS and ground water (70–72). As of the date of this writing, there have been no field tests which have successfully demonstrated nearly complete remediation of a contaminated site using direct food additive status surfactants, but laboratory results on deliberately contaminated soil columns are highly encouraging, especially for microemulsion-based systems. Further, in at least one system, it was found that conditions that produced a middle phase microemulsion with the DNAPL contaminant also dramatically reduced the adsorption of the surfactant (73).

In only a few years, the first major field tests of SEAR will have been completed and researchers will have a much better idea of the viability of the technology. Surfactants certainly exist which are adequate to produce an efficient removal of DNAPLs from the subsurface; however, even more effective molecules may yet be produced. The issue of regulatory acceptance of the injection of surfactants into ground water is still being resolved, but appears favorable as of this date, especially when surfactants with direct food additive status can be used. Much more uncertain is the mode of application. The more cautious lean toward using only solubilization as a removal mechanism and avoiding ultralow interfacial tensions. Those more concerned about the viability of the economics of the process lean toward mobilization. It is almost certain that a field test will be performed using a polymer/surfactant mixture to enhance the ability to control the flow of the injected surfactant solution. Another critical factor which is not addressed in this paper, but which may be the final word on the entire technology, involves the questions of how much knowledge of the subsurface geology is required for a SEAR project to be successful and whether the technology can be developed to obtain that knowledge in a timely and economic manner.

Acknowledgments

Financial support for this work was provided by National Science Foundation Grant CBT 8814147, an Applied Research Grant from the Oklahoma Center for the Advancement of Science and Technology, the Center for Waste Reduction Technologies of the American Institute of Chemical Engineers, Agreement No. N12-N10, and the TAPPI Foundation. In addition, support was received from the industrial sponsors of the Institute for Applied

Surfactant Research including Akzo Nobel, Amway, Colgate-Palmolive, Dow, Dowelanco, DuPont, Henkel, ICI, Kerr-McGee, Lever Reckitt and Colman, Lubrizol, Nikko, Phillips Petroleum, Pilot Chemical, Shell, Sun, and Witco. Dr. Scamehorn holds the Asahi Glass Chair and Dr. Harwell holds the Conoco-DuPont Professorship in chemical engineering at the University of Oklahoma.

References

1. Dunn, R.O., J.F. Scamehorn, and S.D. Christian, Use of Micellar-Enhanced Ultrafiltration to Remove Dissolved Organics from Aqueous Streams, *Sep. Sci. Technol. 20:*257–284 (1985).
2. Dunn, R.O., J.F. Scamehorn, and S.D. Christian, Concentration Polarization Effects in the Use of Micellar-Enhanced Ultrafiltration to Remove Dissolved Organic Pollutants from Wastewater, *Sep. Sci. Technol. 22:*763–789 (1987).
3. Gibbs, L.L., J.F. Scamehorn, and S.D. Christian, Removal of n-Alcohols from Aqueous Streams Using Micellar-Enhanced Ultrafiltration, *J. Membrane Sci. 30:*67–74 (1987).
4. Smith, G.A., S.D. Christian, E.E. Tucker, and J.F. Scamehorn, Equilibrium Solubilization of Benzene in Micellar Systems and Micellar-Enhanced Ultrafiltration of Aqueous Solutions of Benzene, *ACS Symp. Ser. 342:*184–198 (1987).
5. Bhat, S.N., G.A. Smith, E.E. Tucker, S.D. Christian, W. Smith, and J.F. Scamehorn, Solubilization of Cresols by 1-Hexadecylpyridinium Chloride Micelles and Removal of Cresols from Aqueous Streams by Micellar-Enhanced Ultrafiltration, *Ind. Eng. Chem. Res. 26:*1217–1222 (1987).
6. Scamehorn, J.F., and J.H. Harwell, in *Surfactants in Chemical/Process Engineering,* edited by D.T. Wasan, D.O. Shah, and M.E. Ginn, Marcel Dekker, New York, 1988, pp. 77–125.
7. Scamehorn, J.F., and J.H. Harwell, in *Surfactants in Emerging Technologies,* edited by M.J. Rosen, Marcel Dekker, New York, 1987, pp. 169–185.
8. Christian, S.D., and J.F. Scamehorn, in *Surfactant-Based Separation Processes,* edited by J.F. Scamehorn, and J.H. Harwell, Marcel Dekker, New York, 1989, pp. 3–28.
9. Harwell, J.H., and J.F. Scamehorn, in *Management of Hazardous and Toxic Wastes in the Process Industries,* edited by S.T. Kolaczkowski, and B.D. Crittenden, Elsevier, London, 1987, pp. 352–361.
10. Dunn, R.O., J.F. Scamehorn, and S.D. Christian, Simultaneous Removal of Dissolved Organics and Divalent Metal Cations from Water Using Micellar-Enhanced Ultrafiltration, *Colloid. Surf. 35:*49–56 (1989).
11. Christian, S.D., E.E. Tucker, and J.F. Scamehorn, Colloid-Enhanced Ultrafiltration Processes for Purifying Aqueous Streams and Reservoirs, *Am. Environ. Lab.* 13 (February 1990).
12. Leung, P.S., in *Ultrafiltration Membranes and Applications,* edited by A.R. Cooper, Plenum Press, New York, 1979, pp. 415–421.
13. Kandori, K., and R.S. Schechter, Selection of Surfactants for Micellar-Enhanced Ultrafiltration, *Sep. Sci. Technol. 25:*83–108 (1990).
14. Roberts, B.R., A Study of Surfactant-Based Separation Processes, Ph.D. Thesis, University of Oklahoma, Norman, 1993.
15. Scamehorn, J.F., R.T. Ellington, S.D. Christian, B.W. Penney, R.O. Dunn, and S.N. Bhat, Removal of Multivalent Metal Cations from Water Using Micellar-Enhanced Ultrafiltration, *AICHE Symp. Ser. 250:*48–58 (1986).

16. Christian, S.D., S.N. Bhat, E.E. Tucker, J.F. Scamehorn, and D.A. El-Sayed, Micellar-Enhanced Ultrafiltration of Chromate Anion from Aqueous Streams, *AICHE J. 34:* 189–194(1988).
17. Scamehorn, J.F., S.D. Christian, and R.T. Ellington, in *Surfactant-Based Separation Processes,* edited by J.F. Scamehorn, and J.H. Harwell, Marcel Dekker, New York, 1989, pp. 29–51.
18. Scamehorn, J.F., and S.D. Christian, in *Proceedings of the 9th AESF/EPA Conference on Environmental Control for the Metal Finishing Industry,* Chapter N, 1988.
19. Klepac, J., D.L. Simmons, R.W. Taylor, J.F. Scamehorn, and S.D. Christian, Use of Ligand-Modified Micellar-Enhanced Ultrafiltration in the Selective Removal of Metal Ions from Water, *Sep. Sci. Technol. 26:*165–173 (1991).
20. Dharmawardana, U.R., S.D. Christian, R.W. Taylor, and J.F. Scamehorn, Use of Ligand-Modified Micellar-Enhanced Ultrafiltration in the Selective Removal of Metal Ions from Water, *Langmuir 8:*414–419 (1992).
21. Tucker, E.E., S.D. Christian, J.F. Scamehorn, H. Uchiyama, and W. Guo, Removal of Chromate from Aqueous Streams by Ultrafiltration and Precipitation, *ACS Symp. Ser. 491:*86–98 (1992).
22. Christian, S.D., J.F. Scamehorn, E.E. Tucker, E.A. O'Rear, and J.H. Harwell, in *Handbook on Removal of Heavy Metals from Industrial Wastewater,* edited by R.W. Peters, CRC Press, Boca Raton, in press.
23. Simmons, D.L., A.L. Schovanec, J.F. Scamehorn, S.D. Christian, and R.W. Taylor, Ligand-Modified Micellar-Enhanced Ultrafiltration for Metal Ion Separations; Use of N-Alkyltriamines, *ACS Symp. Ser. 509:*180–193 (1992).
24. Scamehorn, J.F., S.D. Christian, D.A. El-Sayed, and H. Uchiyama, Removal of Divalent Metal Cations and their Mixtures from Aqueous Streams Using Micellar-Enhanced Ultrafiltration, *Sep. Sci. Technol. 29:*809–830 (1994).
25. Christian, S.D., and J.F. Scamehorn, editors, *Solubilization in Surfactant Aggregates,* Marcel Dekker, New York, 1995.
26. Nguyen, C.M., S.D. Christian, and J.F. Scamehorn, Solubilization Isotherms for Organic Solutes in Aqueous Micellar Solutions, *Tenside Surf. Deterg. 25:*328–336 (1988).
27. Smith, G.A., S.D. Christian, E.E. Tucker, and J.F. Scamehorn, Solubilization of Hydrocarbons by Surfactant Micelles and Mixed Micelles, *J. Colloid Interface Sci. 130:*254–265 (1989).
28. Mahmoud, F., W. Higazy, S.D. Christian, E.E. Tucker, and A.A. Taha, Solubilization of Hydrocarbons by Surfactant Micelles and Effect of Added Salt, *J. Colloid Interface Sci. 131:*96–102 (1989).
29. Christian, S.D., E.E. Tucker, G.A. Smith, and D.S. Bushong, Calculation of Organic Solute and Surfactant Activities from Solubilization Data, *J. Colloid Interface Sci. 113:*439–448 (1986).
30. Tucker, E.E., and S.D. Christian, Solubilization of Benzene by Aqueous Sodium Octylsulfate—Effect of Added Sodium Chloride, *J. Colloid Interface Sci. 104:*562–568 (1985).
31. Tucker, E.E., and S.D. Christian, Precise Vapour-Pressure Measurements of the Solubilization of Benzene by Aqueous Sodium Octylsulphate Solutions, *Faraday Symp. Chem. Soc. 17:*11–24 (1982).
32. Smith, G.A., S.D. Christian, E.E. Tucker, and J.F. Scamehorn, Solubilization of Hydrocarbons by Surfactant Micelles and Mixed Micelles, *J. Colloid Interface Sci. 130:*254–265 (1989).

33. Tucker, E.E., in *Solubilization in Surfactant Aggregates,* edited by S.D. Christian, and J.F. Scamehorn, Marcel Dekker, New York, 1995, pp. 429–453.
34. Nguyen, C.M., J.F. Scamehorn, and S.D. Christian, Solubilization of n-Hexanol in Mixed Micelles, *Colloid. Surf. 30:*335–344 (1988).
35. Smith, G.A., S.D. Christian, E.E. Tucker, and J.F. Scamehorn, Use of the Semi-Equilibrium Dialysis Method in Studying the Solubilization of Organic Compounds in Surfactant Micelles. System n-Hexadecylpyridinium Chloride Phenol Water, *J. Solution Chem. 15:*519–529 (1986).
36. Christian, S.D., G.A. Smith, E.E. Tucker, and J.F. Scamehorn, Semi-Equilibrium Dialysis: A New Method for Measuring the Solubilization of Organic Solutes by Aqueous Surfactant Solutions, *Langmuir 1:*564–567 (1985).
37. Lee, B.H., S.D. Christian, E.E. Tucker, and J.F. Scamehorn, Effects of an Anionic Polyelectrolyte on the Solubilization of Mono- and Dichlorophenols by Aqueous Solutions of n-Hexadecylpyridinium Chloride, *Langmuir 7:*1332–1336 (1991).
38. Mahmoud, F.Z., S.D. Christian, E.E. Tucker, and J.F. Scamehorn, Semi-Equilibrium Dialysis Study of the Solubilization of Benzoate Anion by Aqueous Hexadecylpyridinium Chloride, *J. Phys. Chem. 93:*5903–5908 (1989).
39. Lee, B.-H., S.D. Christian, E.E. Tucker, and J.F. Scamehorn, Substituent Group Effects on the Solubilization of Polar Aromatic Solutes (Phenols, Anilines, and Benzaldehydes) by N-Hexadecylpyridinium Chloride, *J. Phys. Chem. 95:*360–365 (1991).
40. Uchiyama, H, S.D. Christian, J.F. Scamehorn, M. Abe, and K. Ogino, Solubilization of 2-Phenylethanol by Dodecyldimethylamine Oxide in Aqueous Solution, *Langmuir 7:*95–100 (1991).
41. Lee, B.H., S.D. Christian, E.E. Tucker, and J.F. Scamehorn, Solubilization of Mono- and Di-Chlorophenols by Hexadecylpyridinium Chloride Micelles; Effects of Substituents, *Langmuir 6:*230–235 (1990).
42. Uchiyama, H., S.D. Christian, E.E. Tucker and J.F. Scamehorn, A Modified Semi-Equilibrium Dialysis Method for Studying Solubilization in Surfactant Micelles—Testing the Semi-Equilibrium Assumption, *J. Phys. Chem. 97:*10868–10871 (1993).
43. Christian, S.D., E.E. Tucker, J.F. Scamehorn, and H. Uchiyama, On the Interpretation of Solubilization Results Obtained from Semi-Equilibrium Dialysis Experiments, *Colloid Polym. Sci., 271:*745–754 (1994).
44. Dunaway, C.S., S.D. Christian, and J.F. Scamehorn, in *Solubilization in Surfactant Aggregates,* edited by S.D. Christian, and J.F. Scamehorn, Marcel Dekker, New York, 1995, pp. 3–31.
45. Smith, G.A., S.D. Christian, E.E. Tucker, and J.F. Scamehorn, Group Contribution Model for Predicting the Solubilization of Organic Solutes by Surfactant Micelles, *Langmuir 3:*598–599 (1987).
46. Oosawa, F., *Polyelectrolytes,* Marcel Dekker, New York, 1971.
47. Scamehorn, J.F., S.D. Christian, E.E. Tucker, and B.I. Tan, Concentration Polarization in Polyelectrolyte-Enhanced Ultrafiltration, *Colloid. Surf. 49:*259–267 (1990).
48. Uchiyama, H., S.D. Christian, E.E. Tucker, and J.F. Scamehorn, Solubilization of Trichloroethylene by Polyelectrolyte/Surfactant Complexes in Aqueous Solution, *AICHE J. 40:*1969–1975 (1994).
49. Tharapiwattananon, N., J.F. Scamehorn, S. Osuwan, J.H. Harwell, and K.J. Haller, Surfactant Recovery from Water Using Foam Fractionation, *Sep. Sci. Technol., 31:*1233–1258 (1996).

50. Yin, Y., J.F. Scamehorn, and S.D. Christian, *ACS Symp. Ser. 594:*231–248 (1995).
51. Sabatini, D.A., R.C. Knox, and J.H. Harwell, editors, *Surfactant-Enhanced Subsurface Remediation: Emerging Technologies,* American Chemical Society, Washington, 1995.
52. Nayyar, S.P., D.A. Sabatini, and J.H. Harwell, Surfactant Adsolubilization and Modified Admicellar Sorption of Nonpolar, Polar, and Ionizable Organic Contaminants, *Environmental Sci. Technol. 28:*1874–1881 (1994).
53. Palmer, C.D., and W. Fish, *Chemical Enhancements to Pump-and-Treat Remediation,* U.S. Environmental Protection Agency, EPA/540/s-92/001, Ada, Oklahoma, 1992.
54. Harwell, J.H., R.S. Schechter, and W.H. Wade, Surfactant Chromatographic Movement: An Experimental Study, *AICHE J. 31:*415–426 (1985).
55. Harwell, J.H., F.G. Helfferich, and R.S. Schechter, Effects of Mixed Micelle Formation on the Chromatographic Movement of Surfactant Mixtures, *AICHE J. 28:*448–459 (1982).
56. Pope, G.A., and W.H. Wade, Lessons from Enhanced Oil Recovery Research for Surfactant-Enhanced Aquifer Remediation, *ACS Symp. Ser. 594:*142–160 (1995).
57. Fountain, J.C., C. Waddell-Sheets, A. Lagowski, C. Taylor, D. Frazier, and M. Byrne, Enhanced Removal of Dense Nonaqueous-Phase Liquids Using Surfactants: Capabilities and Limitations from Field Trials, *ACS Symp. Ser. 594:*177–190 (1995).
58. Bourrel, M., and R.S. Schechter, *Microemulsions and Related Systems,* Marcel Dekker, New York, 1988.
59. Shiau, B.J., J.D. Rouse, T.S. Soerens, D.A. Sabatini, and J.H. Harwell, Surfactant Selection for Optimizing Surfactant-Enhanced Subsurface Remediation, *ACS Symp. Ser. 594:*65–81 (1995).
60. Clunie, J.S., and B.T. Ingram, in *Adsorption from Solution at the Solid/Liquid Interface,* edited by G.D. Parfitt, and C.H. Rochester, Academic Press, London, 1983, pp. 105–152.
61. Hough, D.B., and H.M. Rendall, in *Adsorption from Solution at the Solid/Liquid Interface,* edited by G.D. Parfitt, and C.H. Rochester, Academic Press, London, 1983, pp. 247–320.
62. Bitting, D., and J.H. Harwell, The Effects of Counterions on Surfactant Surface Aggregates at the Alumina/Aqueous Solution Interface, *Langmuir 3:*500–511 (1987).
63. Krebbs-Yuill, B., J.H. Harwell, D.A. Sabatini, and R.C. Knox, Economic Considerations in Surfactant-Enhanced Pump-and-Treat, *ACS Symp. Ser. 594:*265–279 (1995).
64. Rouse, J.D., D.A. Sabatini, and J.H. Harwell, Minimizing Surfactant Losses Using Twin-Head Anionic Surfactants in Subsurface Remediation, *Environmental Sci. Technol. 27:*2072–2078 (1993).
65. Rouse, J.D., D.A. Sabatini, J.M. Suflita, and J.H. Harwell, Influence of Anionic Surfactants on Bioremediation of Hydrocarbons, *ACS Symp. Ser. 594:*124–141 (1995).
66. Rouse, J.D., D.A. Sabatini, R.E. Brown, and J.H. Harwell, Evaluation of Ethoxylated Alkylsulfate Surfactants for Use in Subsurface Remediation, *Water Environment Research, 68:*162–168 (1996).
67. Tabatabai, A., M.V. Gonzalez, J.H. Harwell, and J.F. Scamehorn, Optimum Surfactant Selection for Carbonate Reservoirs, *Soc. Pet. Engr. J. 8:*117–122 (1993).
68. Scamehorn, J.F., and J.H. Harwell, in *Mixed Surfactant Systems,* edited by K. Ogino, and M. Abe, Marcel Dekker, New York, 1993, pp. 283–316.
69. Rouse, J.D., D.A. Sabatini, J.M. Suflita, and J.H. Harwell, Influence of Surfactants on Microbial Degradation of Organic Compounds, *Critical Rev. in Environmental Sci. Technol. 24:*325–370 (1994).

70. Shiau, B.J., D.A. Sabatini, J.H. Harwell, and D.Q. Vu, Middle Phase Microemulsions of Mixed Chlorinated Solvents Using Food Grade (Edible) Surfactants, *Environmental Sci. Technol., 30*:97–103 (1996).
71. Shiau, B.J., D.A. Sabatini, and J.H. Harwell, Properties of Food Grade Surfactants Affecting Subsurface Remediation of Chlorinated Solvents, *Environmental Sci. Technol. 29*:2929–2935 (1995).
72. Shiau, B.J., D.A. Sabatini, and J.H. Harwell, Solubilization and Microemulsification of DNAPLS Using Direct Food Additive (Edible) Surfactants, *Ground Water 32:* 561–569 (1994).
73. Shiau, B.J., D.A. Sabatini, and J.H. Harwell, J. *Environmental Engineering, ASCE,* in review.

Chapter 10

New Chelating Agents for the Detergent and Cleaning Industry

Werner W. Bertleff

BASF Aktiengesellschaft, Marketing Specialty Chemicals I, 67056 Ludwigshafen, Germany

Markets and Uses

Metal ions are a problem in virtually all industrial processes that involve water. Different fields such as laundry detergents, household cleaners, power generation, and cosmetics, are among the very diverse sectors of industry and products affected. One of the most common causes of difficulty in industrial processes is the scale formed by precipitation of alkaline earth and heavy metal salts. Thus, pipes can quickly become blocked and valves can seize up. Some heavy metal ions catalyze degradation which can make substances or formulations useless. There are many ways in which such problems can be overcome, at least in principle. In practice, masking ions by chemical means has proved to be the most effective solution to the problem. Such chemical agents must be very stable and able to resist acids and alkalis, as well as oxidizing and reducing agents in the various formulations in which they are employed. They must also possess thermal stability.

For industrial use, only a few classes of chemical substances have proven to fulfill at least the majority of requirements (Fig. 10.1). These include: polyphosphates [e.g., pentasodium tripolyphosphate (STP), and potassium diphosphate]; aminocarboxylates [e.g., nitrilo triacetate (NTA), ethylene diamine tetraacetate (EDTA), diethylene triamine pentaacetate (DTPA), and hydroxyethyl ethylene diamine triacetate (HEDTA)]; hydroxycarboxylates (e.g., citric acid, gluconic acid, tataric acid, and oxidized carbohydrates); and phosphonates [e.g., amino trimethylene phosphonate (ATMP), ethylene diamine tetramethylene phosphonate (EDTMP), and hydroxyethane diphosphonate (HEDP)].

Chelating Agents—Fields of Technical Application

Detergents
Soap Manufacture
Water Treatment
Textile Industry
Printing Ink
Cosmetics and Toiletry
Pulp and Paper Industry
Pharmaceutical Industry
Medical Therapy

Rubber Industry
Photo Industry
Metalplating
Metal Pretreatment
Metal Surface Cleaning
Oil Production
Plant Nutrition & Protection
Environmental Protection

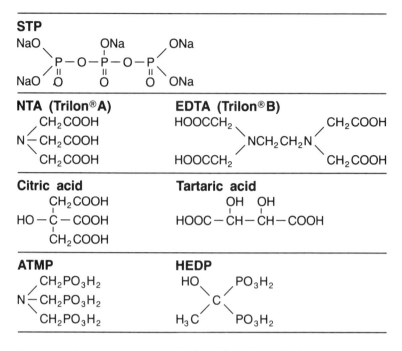

Fig. 10.1. Chemical structures of industrially relevant chelating agents.

Although use of chelating agents is truly universal, some of the chelant classes are more or less restricted in their applicability. The traditional phosphates are good sequestrants for alkaline earth metals. This is why they were the preferred builders in detergent powders. Because they were deemed to contribute to eutrophication of surface waters, they were widely replaced by zeolite/polycarboxylate builder systems in laundry detergents in the industrialized countries. They are still widely used in cleaners, especially in automatic dishwashing detergents.

Aminocarboxylates show good complexing capability for nearly all metal ions. This makes their utility nearly universal. They are chemically stable against oxidation and reduction, insensitive to acids and bases, and easy to formulate for industrial applications. The only drawback of this class of chelants is that not all of them are easily biodegraded. NTA is readily degraded, whereas EDTA, DTPA, and HEDTA are not.

Biodegradability is the main reason why hydroxycarboxylates such as citric acid successfully found their way into various applications. Usually, their complexing power is only medium to low. They are found in detergents, dishwashing cleaners, and particularly in most acidic cleaners.

The phosphonates exhibit a chelating power very similar to that of the aminocarboxylates. Their major disadvantage is their poor biodegradability. Accordingly, their main applications are in the following industries: water treatment in closed loops, in pulp, in paper and in heavy-duty detergents. In household detergents with bleach, they are employed in low concentrations to stabilize the bleach system against traces of heavy metal ions.

Chelating agents of the aminocarboxylate type have been employed industrially for more than 60 years. The worldwide consumption in 1994 is estimated to be 166,000 metric tons, calculated as 100% acid. The greatest consumption is found in North America with 55% followed by Europe with 30%. EDTA is the biggest individual product and accounts for close to two thirds of total consumption. NTA follows with roughly 30% and the rest, including DTPA, play only a minor role.

Physical Characterization

Metal complexes are usually characterized by their stability constants. The definition of the stability constant is as follows:

$$K_{MeZ} = \frac{[Me\,Z^{(m-n)}]}{[Me^{n+}]\,[Z^{m-}]}$$

where K_{MeZ} is the stability constant of chelate; $[MeZ^{(m-n)}]$ is the concentration of chelate; $[Me^{n+}]$ is the concentration of metal ion; and $[Z^{m-}]$ is the concentration of ligand anion. The higher the value, the higher the share at equilibrium of complexed metal. Thus precipitation of uncomplexed metal salts is inhibited. Under defined conditions (for example, 0.1 molar electrolyte concentration and ambient temperature), weak and strong binding ligands for a particular metal ion can be distinguished by their stability constants. Reference (1) contains the most comprehensive data collection of stability constants. As pointed out before, the values for STP and hydroxycarboxylates are rather low, whereas amino carboxylates and phosphonates show a high binding tendency (Table 10.1).

In industrial applications, something other than standardized conditions usually exists. For a reliable assessment of complex formation under the prevailing conditions, competing equilibria have to be considered. The so-called conditional stability constant (2,3) accounts for the stepwise dissociation of polybasic chelants with increasing pH and the formation of metal hydroxo complexes with increasing pH as follows:

$$K_{MeZ}^{cond} = \frac{K_{MeZ}}{\alpha_Z \bullet \alpha_{Me}}$$

where

$$\alpha_Z = \frac{[H_m Z] + [H_{m-1} Z^-] + \ldots + [Z^{m-}]}{[Z^{m-}]}$$

and

TABLE 10.1 Stability Constants of Selected Metal Chelates (ionic strength: 0.1 M; 25°C, log K_{MeZ})

	Ca^{2+}	Mg^{2+}	Fe^{3+}	Cu^{2+}	Mn^{2+}
STP[a]	3.5	3.3	—	—	3.6
DTPA[b]	10.6	9.3	27.3	21.4	15.5
EDTA[b]	10.6	8.8	25.1	18.8	13.8
HEDTA[b]	8.1	7.0	19.8	18.8	10.8
MGDA[b]	7.0	5.8	16.5	—	—
NTA[b]	6.5	5.5	16.3	13.1	7.5
Citric acid[a]	3.5	3.2	11.5	5.9	4.1
Tartaric acid[a]	1.8	—	6.5	3.4	2.5
EDTMP[a]	9.3	8.6	—	21.7	12.7
ATMP[a]	6.7	6.5	—	—	—
HEDP[a]	6.0	6.6	16.2	12.5	9.2

[a]Source: Ref. (1).
[b]Source: BASF.

$$\alpha_{Me} = \frac{[Me^{n+}] + [Me(OH)^{(n-1)}] + \ldots}{[Me^{n+}]}$$

The standard stability constant is corrected by the so-called activity coefficients: α. α corrects for decreased availability of uncomplexed Z and Me due to competing equilibria. In other words, α is a measure of the extent of side reactions. Some authors prefer to use the inverse of α and call it β because its meaning is simpler: it is the fraction of reagent not bound to the target molecule and not involved in side reactions. If the metal ion (Me) in a system of metal ion/chelant is involved in side reactions, then $\alpha_{Me} > 1$. The same holds true for the chelating agent, (Z). Because the aminocarboxylates are polybasic, they undergo stepwise dissociation depending on the pH. At low pH, a considerable number of chelating sites remain undissociated. This means that only a part of the uncomplexed chelate is actually accessible for complexation. Consequently, the α_z values are high at a low pH. With increasing pH, more and more chelant is dissociated. At a pH of 10 or higher, the chelants are essentially fully dissociated. That means that they contribute fully to complexation and no correction is needed. Accordingly, the α_z values fall to zero. In Table 10.2, the log α_z values for various aminocarboxylates are shown as a function of pH.

The concentration of free ligand is dependent on the pH, and the same is true for the concentration of free (i.e., hydrated) metal ions. However, the dependency is inverse to that of the ligands. At a high pH, the metal ions exist partially as more or less stable hydroxo complexes and are less available for the complexing reaction. This explains why the high α_{Me} values at high pH fall to 1 (resp. the log α_{Me} to 0) with decreasing alkalinity. In Table 10.3, the log α_{Me} values for different di- and trivalent

TABLE 10.2 Log α_z as a Function of pH

pH	log α_{NTA}	log α_{EDTA}	log α_{DTPA}	log α_{HEDTA}
1	11.40	17.40	23.50	15.00
3	7.00	10.80	14.90	9.40
5	4.80	6.60	9.40	5.40
7	2.80	3.40	5.30	2.80
9	0.87	1.40	1.70	0.87
10	0.22	0.50	0.68	0.22
11	0.03	0.09	0.14	0.03
12	0.00	0.01	0.02	0.00
13	0.00	0.00	0.00	0.00
14	0.00	0.00	0.00	0.00

Source: Ref. (2).

metal ions are shown. With these α coefficients, the conditional stability constant can be calculated from the standard stability constant. Figure 10.2 illustrates the pH dependent stability of Fe(III) complexes with various aminocarboxylates, whereas Fig. 10.3 shows the stability of the EDTA complexes of different metal ions.

It becomes clear that one must always bear in mind the conditions of application when looking for a suitable ligand. Thus, for masking Fe(III) ions, EDTA is preferable to DTPA up to a pH of 5, whereas the latter is superior at higher pH. Furthermore, the conditional stability constant demonstrates that there is a distinct stability maximum on the pH scale for any individual complex.

TABLE 10.3 Log α_{Me} as a Function of pH

pH	log α_{Mg}	log α_{Ca}	log α_{Cu}	log α_{Zn}	log $\alpha_{Fe(II)}$	log $\alpha_{Fe(III)}$
1	0.00	0.00	0.00	0.00	0.00	0.00
3	0.00	0.00	0.00	0.00	0.00	0.40
5	0.00	0.00	0.00	0.00	0.00	3.71
7	0.00	0.00	0.04	0.00	0.00	7.70
9	0.00	0.00	1.04	0.18	0.12	11.70
10	0.02	0.00	2.00	2.41	0.62	13.70
11	0.15	0.01	3.00	5.41	1.51	15.70
12	0.70	0.08	4.00	8.45	2.50	17.70
13	1.61	0.48	5.00	11.75	3.50	19.70
14	2.60	1.32	6.00	15.53	4.50	21.70

Source: Ref. (2).

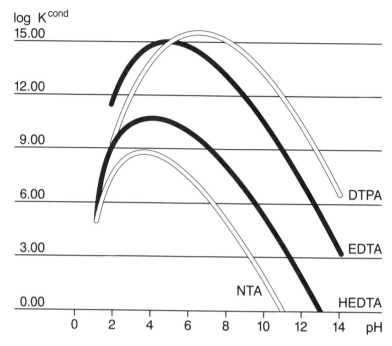

Fig. 10.2. Conditional stability constant for Fe(III) chelates.

Development of New Chelating Agents

As previously mentioned, contrasting the chelating power on the one hand and environmental acceptability on the other, development targets for new chelating agents are clear: their complexing capability should be as high as that of EDTA; they should be biodegradable and nontoxic to humans and marine organisms; raw materials should be readily available and cheap, and the production process should be cost effective. However, the choice of tools for designing new chelants is rather limited. In principle, it is restricted to the selection of electron-donor atoms and to the tailoring of their electronic and steric environment (4). For selecting the appropriate donor atoms, Pearson's "hard and soft acid base" principle is of some use (5). However, in industrial practice, the choice of donor atoms is restricted to only a handful of elements—oxygen, nitrogen, and phosphorus—and their various oxidation states. Therefore, only minor help could be expected from here. For the design of totally new chelants without analogous structures, the calculation of the stability constant using empirical equations starting from known unidentate ligands proves to be helpful. For instance, the equation of Hancock (6) gives useful indications for the search for nonobvious structures based on polyamine and related ligands. However, industrial chelant development is more targeted to variations of known structures such as EDTA and the like. In this context, molecu-

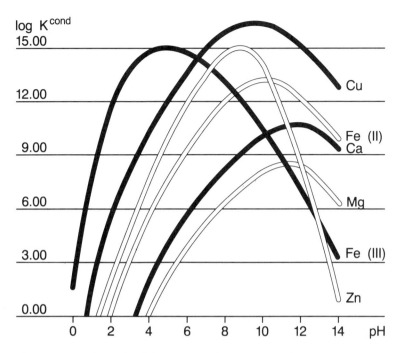

Fig. 10.3. Conditional stability constant for EDTA complexes.

lar modeling has become a valuable tool for the synthetic chemist. The following examples from BASF's laboratories will illustrate the use of modeling today.

As the phrase suggests, complex stability depends on complex phenomena. It is influenced by numerous factors. A high coordination number supports the important chelate effect. On complexation of the metal ion, substitution of water molecules from the metal's hydration sphere occurs, generally giving rise to favorable change in the entropic term of the free-energy expression. Another important factor is the number of rotational degrees of freedom within the ligand. It should be minimal to keep the loss of entropy by complexation low. Another issue is the size of the newly formed chelate ring. An optimum stability is obviously achieved with 5- or 6-member rings. The resulting complex should not carry a residual charge because this would lead to destabilization in a practical multicomponent formulation. Finally, the donor ability of the ligand is dependent on its polarizability, which should be complementary to that of the metal ion.

All of these factors make it very difficult to develop simple theories and strategies. Therefore, we entered all of the available data from a vast number of known calcium complexes into the computer. By that deductive approach, we obtained a substantial amount of information about the preferred constitution of ligands relative to calcium. This allows us to predict the chelating power of new structures to some extent. Many new structures under consideration look quite similar and it would not be easy to predict their chelating power without modeling. The modeling data, for instance, allow us to see whether the ligand exhibits a suitable covering of the central metal. Although both ligands shown in Fig. 10.4 exhibit intense interaction with calcium, the structure on the right side is somewhat overengineered. Its many coordination sites (nitrogen, carboxylate, and oxygen) are favorable from an enthalpic standpoint, but this is obviously unfavorable in an entropic sense. The resulting stability constant decreases, which can be surmised from noting the long distances between the coordinating ligand sites and the central atom.

It is possible, in principle, to design a very good metal complexing agent starting out from the well-known EDTA structure, through modeling using empirical stability correlations. But it turns out to be extraordinarily difficult to make the designed agent environmentally acceptable. Indeed, biodegradability is one of the most challenging development targets for new chelating agents. Little is known about relation between structure and biodegradation. What is known is that there is good correlation of the stability constant and biodegradability: the stronger the chelating power, the worse the biodegradability. This may be due to the complexing agent's binding of physiologically important calcium. For amino carboxylates, seemingly small structural variations lead to considerable changes in biodegradation. But there is a borderline area around log K(Ca) = 7 that provides room for developing new environmentally friendly chelants.

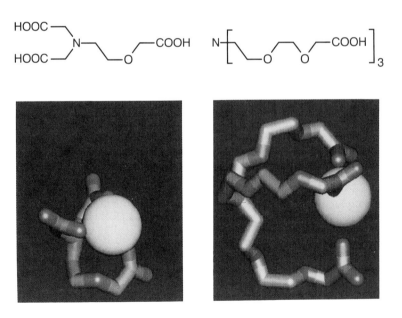

Fig. 10.4. Size of ligand modeling of calcium complexes.

In contrast to screening methods for new pharmaceuticals, where known receptors are the key to new tailor-made substances, we tried to model a suitable receptor known biodegradable chelants would fit to. Beginning with X-ray crystal data for numerous compounds bound to Ca, the molecular volume region common to all biodegradable molecules was calculated. Figure 10.5 shows the "biodegradability boundary" result for EDTA. It can be seen that EDTA has a carboxyl group which extends outside of that volume boundary.

This procedure was used with a high success rate to predict the biodegradability of a large number of molecules. It allows for a fine-tuning of seemingly very similar structures. However, this model is restricted to EDTA-like molecules having significantly lower stability constants, which are close to that of NTA. Extensive "paper" chemistry yielded a promising structure. It is shown in Fig. 10.6 embedded in the "biodegradable volume": methylglycine diacetic acid, MGDA. It is readily biodegradable as can be seen from various ecological test results. In Fig. 10.7, the dissolved organic carbon (DOC) reduction according to the OECD Screening Test 301E is shown.

Synthesis and Properties of Methylglycine Diacetic Acid (MGDA)

With respect to the stability constant and calcium binding capacity, MGDA lies between NTA and EDTA (see Table 10.1). The synthesis of MGDA (Fig. 10.8) can, in principle, start from alanine and go through a classical Strecker-type reaction or

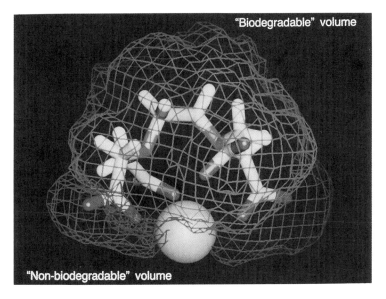

Fig. 10.5. Calcium EDTA relative to biodegradable volume.

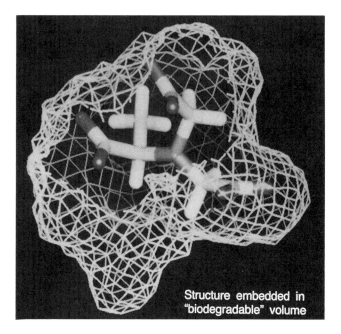

Fig. 10.6. MGDA: A new biodegradable chelant.

alternatively through a double addition of chloroacetic acid. A second possibility is to start from an imino diacetic acid derivative and add acetaldehyde/cyanide again in a Strecker-type reaction. Upon saponification, the sodium salt is obtained.

Application testing with this new compound was extensive, and we have found that MGDA exhibits numerous useful properties in various fields of application. In dairy cleaners, for instance, it contributes to redissolving milk-stone precipitates which consist of different salts, mainly calcium phosphate and fatty acid salts, together with albumen. The redissolving ability especially at 60°C comes close to that of EDTA, which is presently used, and is superior to NTA (Fig. 10.9).

Autodishwashing tests in silicate-free formulations show MGDA to be an effective scale inhibitor for hard water (Fig. 10.10). The results, along with those for NTA, are judged visually and are expressed in "marks". Mark zero means that no scale precipitate was present. The results vary depending on the concentrations employed, but the results with an industrially realistic concentration of about 15% are promising, especially with porcelain, stainless steel, and glass.

MGDA was also tested as a builder substitute in powder laundry detergents. Aminocarboxylates are known to show good dispersing and anti-encrustation effects in detergents. Thus, NTA is employed in institutional and household detergents in several countries. The reduction of builder content in detergents has considerable attractiveness in the light of volume reduction and smaller packages. On the other

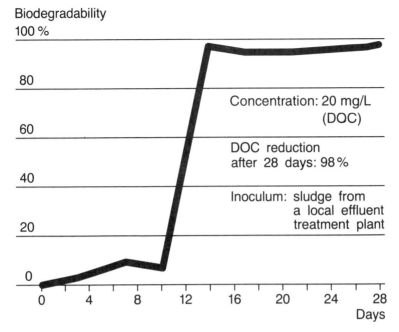

Fig. 10.7. MGDA biodegradation test—modified OECD Screening Test (301E).

hand, merely lowering builder content leads to reduced primary detergency with little or no effect on encrustation (Fig. 10.11). However, compensating 10% lower builder content by adding back 5% MGDA shows beneficial effects: detergent efficiency is restored and encrustation is lowered significantly. The effect of substituting 8% MGDA for 16% builder is even more dramatic.

The performance of MGDA in liquid detergents was also examined. Here it exhibits strong benefits in removing different stains (Fig. 10.12). It equals NTA in removing tea and blood stains. For blood stains in particular, its performance is superior to that of EDTA and the phosphonate EDTMP, each of which is nonbiodegradable.

Summary

The results presented here provide only a rough overview of some of our observations on MGDA. Testing in different applications continues. With MGDA, a new biodegradable complexing agent was developed using a rational approach including structural modeling and modeling of a "biodegradable volume." MGDA exhibits a versatile spectrum of potential uses. It is easily synthesized using raw materials and technical equipment, which are commonly employed in aminocarboxylate manufacturing. The material has promising properties in cleaners for the food and beverage industry,

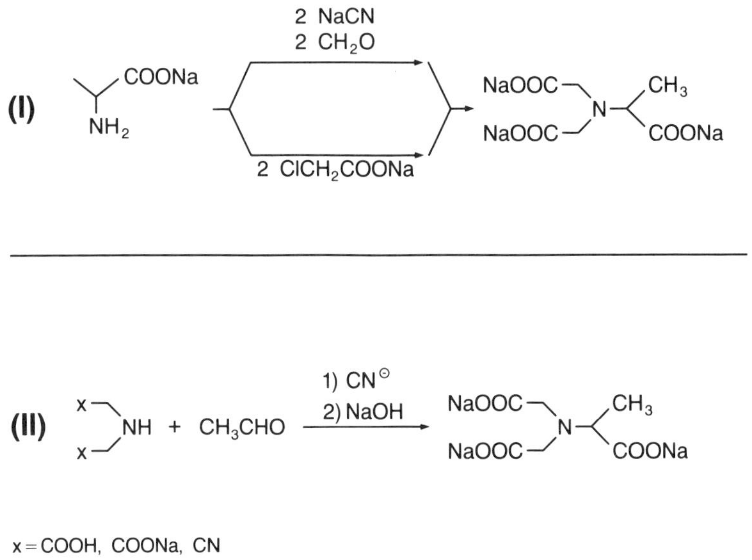

Fig. 10.8. Synthetic routes to methylglycine diacetic acid (MGDA).

in autodishwashing cleaners, and in powder and liquid laundry detergents. It has a unique chelating profile and supplements the assortment of chelating agents currently available. MGDA is not listed on the TSCA or the European EINECS list. Investigations for the listing procedure according to the German Chemical Substances Act are far advanced. The substance has gotten permission to be marketed, meanwhile. MGDA is likely to be offered to customers on an industrial scale by 1997. The global commercialization is strongly influenced by the regulatory listing procedures required in various countries.

References

1. Martell, A.E., and R.M. Smith, *Critical Stability Constants,* Vol. 1–3, Plenum Press, New York, 1974.
2. Ringbom, A., and E. Wänninen, in *Treatise On Analytical Chemistry,* 2nd edn., Part 1, Vol. 2, edited by I.M. Kolthoff, and Ph.J. Elving, John Wiley & Sons, New York, 1979.
3. Ringbom, A., *J. Chem. Educ. 35:*282 (1958).
4. Hancock, R.D., *Chem. Rev. 89:*1875 (1989).
5. Pearson, R.G., *J. Am. Chem. Soc. 85:*3533 (1963).
6. Hancock, R.D., and F. Marsicano, *J. Chem. Soc., Dalton Trans.:*1096 (1976).

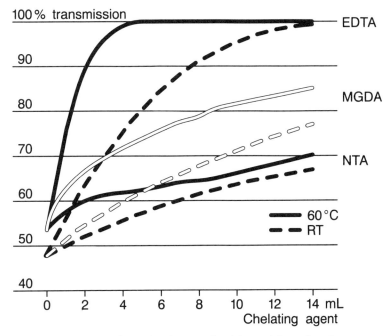

Fig. 10.9. MGDA in dairy cleaning—redissolving calcium phosphate.

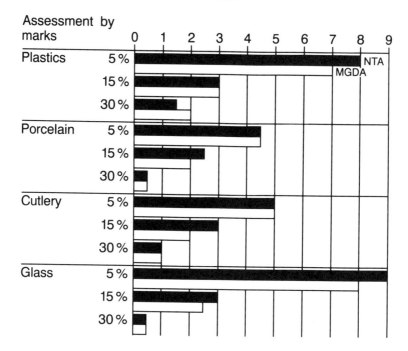

Fig. 10.10. MGDA in autodishwashing—scale inhibition (silicate-free formulation).

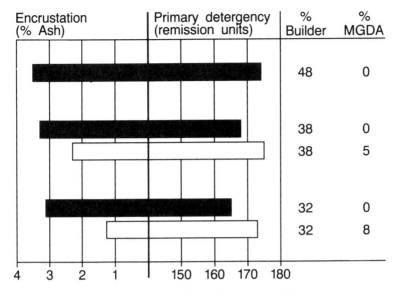

Formulation (major ingredients): 9% FAS, 10% nonionics, 36% zeolite A, 12% soda, 15% perborate, 4% TAED
Conditions: Launder-o-meter, 60°C

Fig. 10.11. MGDA as builder substitute.

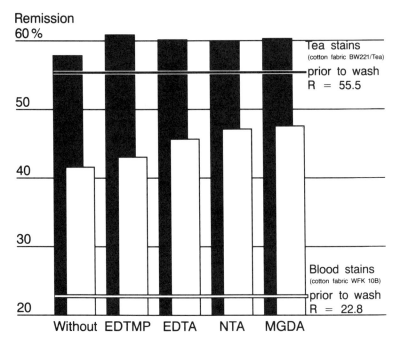

Fig. 10.12. MGDA—stain removal in liquid detergents.

Chapter 11

Mechanism of Enzyme Action and Correlation with Cleaning Performance

Peter F. Plank[a,*], Stephen J. Danko[a], Judy Dauberman[a], Matthew J. Flynn[a], Constance Hsia[a], Deborah S. Winetzky[a], and Edward D. DiCesare[b]

[a]Genencor International, Inc., Palo Alto, CA, and [b]Rochester, NY

Background

The use of industrial enzymes in detergents for improving cleaning performance is increasing. Enzymes currently used include amylases, cellulases, lipases, and proteases. The challenge faced by formulators is selecting the best enzyme(s) from the products available to them. Enzyme activity assays measure the hydrolysis rate of a synthetic substrate, for example, dimethyl casein or a synthetic peptide, such as suc-Ala-Ala-Pro-Phe-p nitroanilide (suc-AAPFpNA). The rate of hydrolysis varies depending on the substrate, the assay conditions, and the enzyme being evaluated.

Protease activity on these synthetic substrates cannot be directly translated into wash performance for enzymes derived from different organisms. For example, under a defined set of assay conditions, a protease derived from *Bacillus licheniformis* has a 16-fold advantage over a protease from *Bacillus lentus* on the synthetic peptide, suc-AAPFpNA. In actual wash performance evaluations, the *B. lentus*-derived protease outperforms the *B. licheniformis*-derived protease. The latter reaches a plateau at a lower enzyme usage level in a dose-efficiency curve (DEC). It requires three times the amount of *B. licheniformis* protease to obtain equivalent wash performance to the *B. lentus* protease. This phenomenon holds true for the other hydrolases as well.

Enzymes have been used by industry for the past 100 years. The detergent industry has actively pursued their use over the past 25 years (1). During this time, enzyme fermentation, recovery, and granulation technology have improved. These improvements addressed many of the safety concerns associated with enzyme usage in detergents in the early 1970s. The detergent industry continues to be a strong proponent of enzyme technology and the use of enzymes in detergent formulations. Product package labelling promotes the use of enzymes for a variety of different applications, all touting their benefits to the consumer. Proteases, the first enzymes to be introduced into detergent formulations, are used to clean difficult-to-remove protein-based stains such as blood, egg, and grass. α-Amylases have been included in detergent formulations to remove starch-based stains such as gravy, pudding, and potato. Lipases demonstrate their cleaning advantages on triglyceride-based stains such as margarine and oil. Most recently, cellulases have been included for color maintenance or restoration benefits on fabric. All of these enzymes are specific for a particular type of stain yet all are classified as hydrolases, based on their mechanism of action.

---
*Author to whom correspondence should be directed.

Table 11.1 depicts enzyme classification (2) as specified by the International Union of Biochemistry (IUB). Enzyme classification is based on their reaction mechanisms.

The focus of this paper will be on hydrolase enzymes used by the detergent industry. In particular, protease, α-amylase, lipase, and cellulase enzymes, their mechanism of action and correlation of enzyme activity with cleaning performance.

TABLE 11.1 Enzyme Classification

IUB Classification	Mechanism of action
Oxidoreductase	Catalyzes oxidation/reduction reactions, e.g., glucose oxidase.
Transferase	Catalyzes transfer of a functional group between molecules.
Hydrolase	Catalyzes the addition of water across a bond, e.g., protease, lipase, α-amylase, and cellulase.
Lyase	Catalyzes the addition of a functional group to a double bond or generates double bonds, e.g., pectin lyase (transeliminase).
Isomerase	Catalyzes the isomerization or rearrangement of a molecule, e.g., glucose isomerase.
Ligase (synthetase)	Catalyzes the breaking or formation of two molecules concomitant with cleavage of a nucleoside triphosphate, e.g., glutamine synthetase.

Amylases

α-Amylases (α-1,4-glucanohydrolases) break down the α-1,4 linkages of high molecular weight (MW) starch polymers resulting in an overall reduction in starch solution viscosity. Starch is a high MW polymer derived from D-glucose units linked by α-1,4 glycosidic bonds, with α-1,6 bonds at branching points. One potential mechanism for this hydrolysis process is depicted in Fig. 11.1. α-Amylase can degrade starch only by hydrolysis of the α-1,4 linkages, generating branched dextrins and oligosaccharides. It is unable to hydrolyze the α-1,6 branch points because this requires an enzyme specific for this type of polymer linkage, a pullulanase or isoamylase. The benefits of adding α-amylases to detergents is their ability to degrade the starch molecule into intermediate molecular weight oligosaccharides or reducing sugars. These lower MW products are more readily removed by the mechanical and chemical action of the washing machine and detergent, respectively.

The various types of α-amylases (3) available are presented in Table 11.2. They differ from one another not only in substrate specificity, but also in pH and temperature optima. Starch hydrolysis reaction rates can be monitored by the measurement of viscosity reduction or by the reaction of nonhydrolyzed starch with iodine. The reaction rates are dependent on the pH, temperature and degree of polymerization of the substrate.

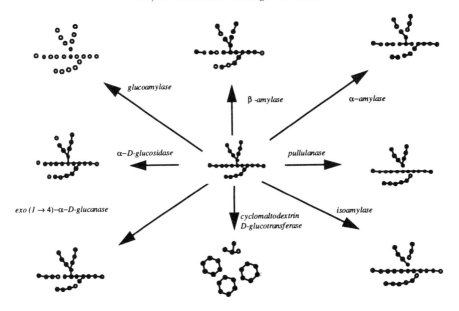

Fig. 11.1. Starch hydrolysis mechanism.

TABLE 11.2 α-Amylase Comparison

Enzyme	Source organism	pH Activity range (optimum)	Temperature range (°C) (inactivation T)
Bacterial α-amylase	*Bacillus subtilis*	4.5–9.0	70–85
	B. amyloliquefaciens	(6.5–7.5)	(95)
Thermostable bacterial α-amylase	*B. licheniformis*	5.8–8.0 (7.0)	90–105 (120)
Fungal α-amylase	*Aspergillus oryzae*	4.0–7.0	55–60
	A. niger	(5.0–6.0)	(80)

α-Amylases typically used in the detergent industry are either bacterial α-amylases (RAPIDASE (4) or BAN (5)) or thermostable bacterial α-amylases (MAXAMYL (4) or TERMAMYL (5)). These bacterial α-amylases have been shown to be calcium dependent. Starch hydrolysis by these α-amylases is increased at elevated calcium concentrations. More recently it has been demonstrated that through the use of genetic engineering, the calcium dependency (6) of an α-amylase can be altered.

Lipases

Lipase enzymes have recently been introduced into detergent products. Detergent manufacturers have touted their efficacy at removing triglyceride-based stains from fabric. Some commercially available lipases (7) are presented in Table 11.3. Lipase enzymes hydrolyze triglycerides only at the lipid-water interface. The hydrolysis mechanism of a triglyceride by a lipase to generate the free fatty acid and glycerol is depicted in Fig. 11.2. The mechanism shown in the figure is reversible, which allows lipases to be used in other applications. The lipase products currently available to the detergent formulator are: LIPOLASE (5) (*Humicola languinosa* lipase expressed in *Aspergillus oryzae*), LIPOMAX (4) (*Pseudomonas* sp.) and LUMAFAST (4) (*Pseudomonas* sp. lipase expressed in *B. subtilis*). New applications for lipases include the hydrolysis of triglycerides, transesterification, stereospecific hydrolysis of racemic esters, and hydrolysis of oily stains in laundry.

TABLE 11.3 Biochemical Properties of Various Lipases

Source (species)	pH range (optimum)	Temperature (°C) (optimum)	pH for stability
Porcine pancreas	6.5–9.5 (7.5–8.5)	40–45	5.5–7.5
Mucor javanicus	5.5–8.0 (7.0)	40–45	4.5–6.5
Pseudomonas sp.	4.0–5.0 (7.0–8.5)	50–60	4.5–10.0
Humicola sp.	7.0–11.0 (9.0–11.0)	50–60 (35–40)	NA

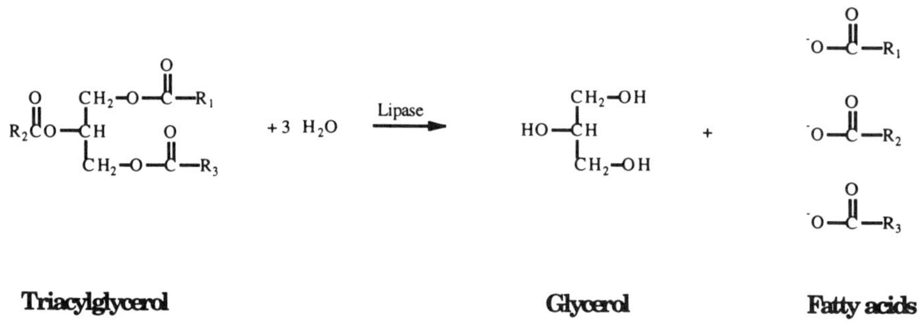

Fig. 11.2. Mechanism of triglyceride hydrolysis.

Cellulases

Cellulase enzymes are the most recent introduction into detergent formulations. Their performance advantages have been touted for color restoration, color maintenance, softening, surface-polishing, and removal of particulate soil. Here again we are faced with a diversity of products derived from different microbial sources for use in detergent cleaning. Cellulases are enzyme complexes which coordinate their component activities in the stepwise decomposition of cellulose and their derivatives to glucose. The three main component activities responsible for this coordinated effort are exocellulase or cellobiohydrolase (CBH), endocellulase or endoglucanase (EG), and β-glucosidase (BG). A potential mechanism for cellulose degradation is depicted in Fig. 11.3. The needs of the detergent industry are for a cellulase that is efficacious at both high pH (7 to 11) and low temperature (<40°C). Table 11.4 lists a number of cellulases of microbial origin only a few of which meet these requirements. The current trend is to isolate the individual cellulase components and evaluate them for efficacy.

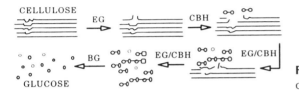

Fig. 11.3. Mechanism of cellulose degradation.

TABLE 11.4 Biochemical Properties of Various Cellulases

Source (species)	pH range (optimum)	Temperature (°C)
Humicola sp.	5.0–8.0 (6.5)	45–50
Trichoderma sp.	3.0–7.0 (5.5)	40–45
Aspergillus sp.	4.0–6.0	40
Bacillus sp.[a]	4.0–12.0 (9.0 to 10.0)	40

[a]*Source:* Ref. (8).

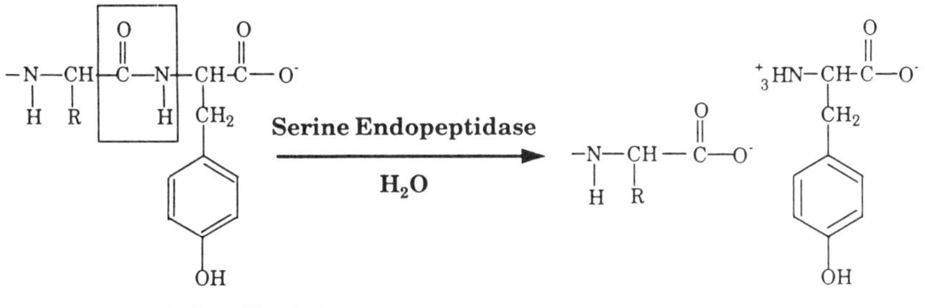

Fig. 11.4. Peptide bond hydrolysis.

Proteases

Protease enzymes were the first hydrolases introduced into detergent formulations specifically for the degradation of protein-based stains. These proteases had to be efficacious over wide pH (7–12) and temperature (10–90°C) ranges. These demands by the industry led to the development of proteases which were bacterial in origin and further classified as alkaline serine proteases. The high pH requirements are a reflection of the commercial detergent formulations found in the marketplace. The overall hydrolysis of a protein by a serine endo-peptidase is depicted in Fig. 11.4.

The serine endopeptidase contains a catalytic triad of amino acids at the active site: an aspartyl residue with a $\beta - COO^-$ as a potential nucleophile, a histidine containing the imidazole group as a functional group, and a serine residue with a $\beta - OH$ as a functional group. The hydrolysis mechanism (9) can be described in more detail as follows. The serine group initiates the nucleophilic attack on the peptide bond to form a tetrahedral intermediate which undergoes an active hydrogen transfer facilitated by both the histidine and aspartyl residues. The net effect is the addition of water across the peptide bond generating two polypeptides smaller than the original protein.

Measurement of Enzyme Activity

Enzymes [10] are characterized by both the type of reaction catalyzed and their catalytic efficiency. Catalytic efficiency indicates the magnitude of an enzyme's effect. When considered on a unit weight basis, catalytic efficiency provides an indication of the enzyme's value. Enzyme activity that delivers a realizable benefit in a specific application is the main concern of the formulator. This quantification of enzyme activity can then be used to determine how much enzyme is required to achieve the desired effect. This forms the basis for comparison of several enzyme products. The activity units specified by the supplier are based on the enzyme's catalytic effect defined under optimal reaction conditions, pH, temperature, ionic strength, and substrate concentration. This is the basis of enzyme kinetics.

The IUB has attempted to standardize enzyme activity. This is admirable in principle, yet overlooks some basic tenets when applied to industrial enzymology. In an industrial application, one must consider the following. First, industrial enzymes do not react with pure, well-defined chemical substrates. They catalyze the conversion of complex mixtures of compounds whose molecular composition is seldom well defined. Second, solubility characteristics of the industrial stain seldom lend themselves to being available at their optimal substrate concentrations. As a direct consequence, substrates encountered under typical application conditions are subject to complicated reaction kinetics. Third, another issue arises when one tries to define standard temperature and optimal conditions. Using a standard temperature (25°C) may be unrealistic, especially if the industrial process conditions are at higher temperatures. pH optima can also be affected by elevated temperatures. This does not even address the effects that substrate modification may have on the pH and temperature optima.

Standards of Activity

The question that ultimately arises in any discussion between customer and supplier is "Why isn't there a standard or universal activity unit used by the enzyme suppliers?" There are two approaches available to address the use of a universal activity unit.

They are based on the use of synthetic substrates as model systems for industrial catalyzed processes under very specific reaction conditions, and the use of standard process conditions which approximate the industrial reaction being catalyzed.

In the first approach, the hydrolysis or digestion of the peptide bond by the protease is critical in the measurement of its activity. A number of spectrophotometric methods for protease enzyme activity measurement are available, yet all involve the hydrolysis of a low molecular weight synthetic peptide, succinyl-alanyl-alanyl-prolyl-phenylalanyl-para-nitroanilide (suc-AAPFpNA) or a chemically modified complex protein substrate, *N,N*-dimethylcasein (DMC), or hemoglobin, the Anson method (11). The information provided by these low MW synthetic peptides may be misleading as models for the natural substrate which is typically a complex, high molecular weight protein. Chemically modified forms of the natural substrate may be a better alternative, yet they still do not adequately model the natural substrate.

The second approach is an extreme approach and fails when one considers that different users of a standard enzyme product may use different processing conditions and different types of substrates. It is therefore impossible to define a standard substrate and conditions which would meet everyone's needs.

The approach used by enzyme suppliers in the industry is derived from the first approach in which the enzyme activity units are defined under a standard set of conditions, substrate, pH, temperature, and ionic strength. The rationale for this approach is that the biochemical reactions used in many enzymatic assays are complex and can be difficult to reproduce between laboratories. The systematic error associated with a given assay result can be influenced by many parameters in each laboratory's assay set-up, including instrumentation, reaction temperature, reaction pH, and ionic strength. An attempt to control and standardize these parameters at the laboratories of various enzyme suppliers would be required to obtain consistent crossover of enzyme activity results. This standardization process is extremely labor intensive, time consuming, and difficult to maintain over the long term.

As a consequence, enzyme suppliers attempt to eliminate any systematic error among intracompany laboratories by calibrating each laboratory's assay method using an enzyme standard. These in-house reference standards and assay methods are typically issued and maintained at one site. The attempt to extend this type of standardization of enzyme calibrators and methods to be shared among enzyme suppliers would encounter a number of practical barriers. To address the customer's perspective, however, specific assay conditions, assay protocols, standardized "reference" samples, and substrates are typically available from the enzyme manufacturers upon request.

The two principal methods currently used in the industry for the measurement of protease enzyme activity incorporate the use of either a synthetic peptide or a chemically modified natural substrate. A general overview (12) of these two enzyme activity protocols, incorporating the synthetic peptide suc-AAPFpNA and the chemically modified substrate dimethyl casein (DMC), is given in Figs. 11.5 and 11.6. The advantages and disadvantages associated with the use of these two substrates for enzyme activity analysis are presented in Table 11.5.

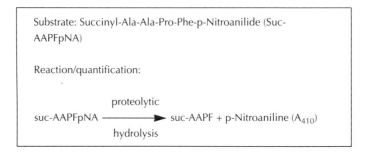

Fig. 11.5. pNA assay method. *Source:* Ref. (12).

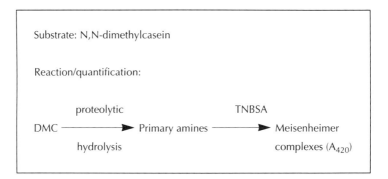

Fig. 11.6. Dimethyl casein (DMC) assay method. *Source:* Ref. (12). TNBSA, trinitrobenzene sulfonic acid.

TABLE 11.5 AAPFpNA vs. DMC Assay[a]

Parameter	pNA	DMC
Time	6 min	25 min
Enzyme Concentration	1–5×10^{-7} Anson Units (AU)/mL	1–5×10^{-6} AU/mL
Interference	None	Primary amines
Precision	<0.5% Relative Standard deviation (RSD)	5% RSD
Advantages	• Quick • Accurate and precise • Stable reagents • Little or no matrix effects • Linear response with time • pH independent	• Little enzyme-to-enzyme variation
Disadvantages	• Significant enzyme-to-enzyme variation	• Time consuming • High blank • Unstable reagents • Primary amine interference

[a]AAPFpNA, Alanyl-alanyl-prolyl-phenylalanyl-para-nitroanilide; DMC, *N,N*-dimethylcasein.

It is apparent that the higher sensitivity and lack of known interferences associated with the pNA assay method offer significant advantages over the DMC method for enzyme activity measurement.

Interpreting Measurements of Activity

The pNA method is an extremely sensitive assay for protease enzymes derived from different microbial sources. This method can therefore be used to help characterize protease enzymes from different microbial origins based on their kinetic constants, k_{cat} and K_M, for different synthetic peptide substrates.

A comparison of the relative rates of hydrolysis for a *Bacillus* sp. alkaline protease, LG-12[13], which exhibits biochemical properties, pH, and temperature optima, similar to those of the commercially available *B. lentus* protease, ESPERASE® (5) is shown in Table 11.6. The synthetic substrate profiles in Table 11.6 indicate that LG-12 is most likely not a *B. lentus*-derived protease, but more likely is related to *B. amyloliquefaciens*, BPN', which is similar in profile to those on the list of synthetic substrates.

TABLE 11.6 Synthetic Substrate Profiles (s-AAPFpNA/s-XXXpNA)

Substrate	B. amyloliquefaciens	B. lentus	LG-12
s-AAPFpNA	1	1	1
s-AAPLpNA	1.8	6	1.5
s-AAPMpNA	3.3	1.5	4.6
s-AAApNA	75	6.5	62.1

To carry this discussion one step further, we can compare the kinetics data, k_{cat}/K_M, based on the rates of hydrolysis of the synthetic peptide, suc-AAPFpNA, by the *B. licheniformis*-derived protease, e.g., MAXATASE® (4) with the *B. lentus*-derived protease, MAXACAL® (4). The Michaelis constant, K_M, is defined roughly as a measure of the binding affinity of the enzyme for the substrate according to Fig. 11.7.

In this case $K_M = k_1/k_{-1}$, when k_{cat} is relatively small compared with k_{-1} ($k_{cat} \ll k_{-1}$). K_M can also be defined as the substrate concentration required to achieve one half of the maximum velocity of the enzymatic reaction as depicted in Fig. 11.8 (Danko, S.; Genencor International Inc., Palo Alto, CA, unpublished results).

Fig. 11.7. Simple enzyme-substrate reaction model.

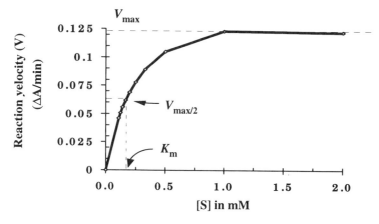

Fig. 11.8. Enzyme kinetics model.

The turnover number, k_{cat}, reflects the number of molecules of substrate converted into product, P, per unit time. When the enzyme is fully saturated with substrate, $[S] \gg [K_M]$, then k_{cat} is directly proportional to the rate of product formation. In terms of enzyme efficiency, it is desirable to have an enzyme which has a high k_{cat}/K_M ratio and a K_M higher than the substrate concentration it encounters in order to avoid saturation of the enzymme with substrate. Industrial applications typically involve $[S] \gg K_M$; therefore k_{cat}/K_M values are of limited use in selecting enzymes targeted for these applications. For suc-AAPFpNA, the k_{cat}/K_M ratio for MAXATASE is approximately 3.8×10^6, while that for MAXACAL is 2.3×10^5. One interpretation of these data is that the rate of hydrolysis of the synthetic substrate, suc-AAPFpNA, for MAXATASE (4) is ~16 times that of MAXACAL (4). A natural conclusion is that, with a higher rate of hydrolysis on the synthetic substrate suc-AAPFpNA, MAXATASE must therefore be the better enzyme. This is useful information to the enzymologist, but how can this be meaningfully translated into useful wash performance information for the detergent formulator? The assumption based on the rate of hydrolysis data would indicate that similar wash performance should be attainable between these two enzymes if we used 16 times more MAXACAL than MAXATASE.

It is readily observed that this is not the case. Fig. 11.9 depicts an enzyme dosage response curve in a commercially available heavy-duty laundry powder (pH 10.5; 1% aqueous solution) using these two enzymes. To achieve similar wash performance between these two enzymes, one would have to use approximately three times as much MAXATASE as MAXACAL. The wash performance and the enzyme activity relationship do not appear to correlate. Why does this apparent disparity between enzyme activity on a synthetic substrate and actual cleaning performance exist? This can be answered as follows: first, enzymatic activity measurements, using the pNA or DMC assay methods, were not developed as representative models for the actual enzymatic cleaning performance, and second, enzyme cleaning performance measures the differences between a heterogeneous and a homogenous reaction system. Protease

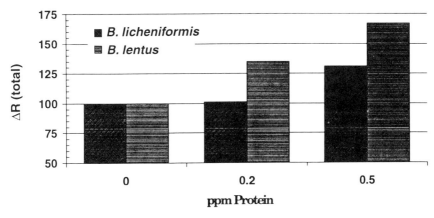

Fig. 11.9. Enzyme dose response evaluation: total cleaning performance. U.S. HDPD 1.5 g/L; enzyme dosage as indicated; terg-o-tometer: 100 rpm t = 15 min wash/10 min rinse; dH = 150 ppm, 2/1 Ca^{2+}/Mg^{2+}; T = 40°C; soil cloths used: EMPA 111, 112, 116 and 117.

enzyme kinetics during the wash, which is a heterogeneous insoluble substrate system, are quite different from the kinetics in a homogenous soluble substrate system.

Figure 11.10 (14) shows the pH profile of SAVINASE (5) and the differences between a homogenous and a heterogeneous system. The key point is that the activity as measured on denatured hemoglobin is relatively flat (100% activity) over the pH range, 8–11. The wash performance over this same pH range using a cotton cloth stained with grass is significantly lower than the enzyme activity until pH 11 is reached. Here again we see that wash performance and enzyme activity as measured by hemoglobin hydrolysis do not seem to show any correlation.

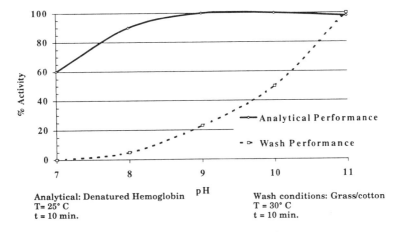

Fig. 11.10. Analytical activity and wash performance profiles for SAVINASE.

How does the formulator determine which enzyme will afford the best performance? Enzymes, which are sold on a cost per unit weight basis, have to be evaluated in an actual wash performance test. Enzyme performance can be assessed in one of three ways: equal activity basis, equal weight or dosage basis, or equal cost basis.

Evaluations based on equal activity are difficult to do for several reasons. First, enzymes from different microbial sources do not necessarily correlate as seen in Fig. 11.9. Second, enzyme suppliers typically use different assay protocols and define their activity units differently. Third, detergent formulators typically do not have the time to do their own analytical determinations of enzyme activity using a standard protocol with the corresponding analytical standards for comparison purposes.

Enzyme performance evaluations carried out on an equal weight or equal dosage basis do not address the question of enzyme cost or relative enzyme activity. This can mislead the formulator into believing that one product being evaluated is better than another product and may result in higher overall costs in the formulation.

Performance evaluations on an equal cost basis are the most representative of enzyme efficacy and directly represent the actual usage costs in the formulation. It should be kept in mind, however, that due to the hyperbolic nature of a dose efficiency curve, a complete dosage efficiency curve is required to make a fair comparison. This evaluation results in a direct correlation between the formulator's enzyme costs per unit activity, which is readily measured as improvements in cleaning performance.

Conclusions

Enzyme activity analysis and its lack of correlation with wash performance is not easy to understand for two reasons: First, the differences in the dynamics of the reaction kinetics between the two systems—homogenous soluble substrate, blood or hemoglobin, vs. heterogeneous complex insoluble substrate, blood stained on fabric, and second, differences observed in the detergent matrices being evaluated by the formulator. In all cases, the only reliable approach to assess enzyme efficacy is to carry out a wash performance evaluation. The advantages of this approach are that direct evaluations can be made between enzymes irrespective of their microbial origin, and direct evaluations of the final formulation costs can be made by the formulator using an unbiased method of comparison.

References

1. Dambman, C., P. Holm, V. Jensen, and M.H. Nielsen, How Enzymes Got into Detergents, *Developments in Industrial Microbiology 12:*11–23 (1971).
2. Walsh, C., *Enzymatic Reaction Mechanisms,* W.H. Freeman and Company, San Francisco, CA, 1979, p. 22.
3. Gerhartz, W., ed., in *Enzymes in Industry, Production and Applications,* VCH Verlaggesellschaft mbH, D-6940 Weinheim, 1990, pp. 77–80.
4. Available from Genencor International Inc., Rochester, NY.
5. Available from Novo-Nordisk Bioindustrial Inc., Bagsvaerd, Denmark.

6. Aehle, W., Improvement of an α-Amylase for Industrial Applications by Protein Engineering, presented at the 86th AOCS Annual Meeting & Expo in San Antonio, TX, May 8, 1995.
7. Gerhartz, W., ed., in *Enzymes in Industry, Production and Applications,* VCH Verlaggesellschaft mbH, D-6940 Weinheim, 1990, p. 89.
8. EPA 270974 A2, Kao Corporation, 1987.
9. Walsh, C., *Enzymatic Reaction Mechanisms,* W.H. Freeman and Company, San Francisco, CA, 1979, pp. 94–97.
10. Godfrey, T., and J. Reichelt, eds., in *Industrial Enzymology: The Application of Enzymes in Industry,* The Nature Press, New York, NY, 1983, Ch. 2.1, pp. 8–40.
11. Anson, M.L., *J. Gen. Physiol. 22:*79–89 (1938).
12. Rothgeb, T.M., B.D. Goodlander, P.H. Garrison, and L.A. Smith, *J. Am. Oil Chem. Soc. 65:*806–810 (1988).
13. Noncommercial samples available from Genencor International Inc., So. San Francisco, CA, for evaluation under biological sampling agreement.
14. Aaslyng, D., E. Gormsen, and H. Malmos, Mechanistic Studies of Proteases and Lipases for the Detergent Industry, *J. Chem. Tech. Biotechnol. 50:*321–330 (1991).

Chapter 12

The Cost of New Product Development in Chemical Specialties

Daniel R. Dutton

    Lonza Inc., Annandale, New Jersey

Background

Sometime around 1990, the way senior management of the chemical industry viewed Research and Development began to change. Maybe it is more appropriate to say that senior management finally began to focus some attention on an asset that required some scrutiny. The writing was really on the wall back in the seventies. One of the first things to be eliminated or cut back in the decades of mergers and acquisitions was Research and Development. For the most part, except for litanies of patents and publications, few research leaders could really quantify the value of Research and Development to the business. Because of this disconnection, the Research and Development management of the U.S. aided and abetted the decline of one of our national treasures, industrial research.

    The following statements are generalizations. There are managers who are effective at selling the business value of a research organization to upper management. They can be proud of the fact that their Research and Development organization is seen by senior management as a key asset with significant measurable value and bottom-line contribution. For managers who have not taken this approach, all is not lost. But be fully aware that the future is almost totally dependent on being able to demonstrate this value. Some of these ideas come from, or have been inspired by, the book by Roussel, Saad, and Erickson entitled *Third Generation R&D* (1). The teachings of this book can serve as a useful model for developing a value-based Research and Development unit. A second valuable resource is the Industrial Research Institute journal *Research Technology Management,* which recently published a twenty-five year index (2).

    There are costs associated with doing research and product development but when you add up all of these costs, they become an investment. The first thing we must do is look at research and product development as an investment. It is our job and responsibility as managers of research to maintain the cost effectiveness of our operations—to minimize essential costs and eliminate unnecessary ones. We must constantly ask "how does this cost contribute to the value?" Everything can be tied to a dollar, and it must be.

Costs

One thing at which we all excel is defining how much we spend on Research and Development. Annual surveys are done by a variety of magazines, universities, and private concerns. The one measure that lags significantly is the measure of value. What did we get for what we spent? Typically, companies are judged or ranked by the size of the R&D budget vs. sales, rather than by profits generated as a direct outcome of R&D investment vs. R&D budget. The profitability of a company is an indirect measure but a poor one at best.

Depending on the size of a company and its commitment to Research and Development, fully loaded costs per man-year range from $90,000 to $165,000. Make sure you understand fully all of the costs for which you are responsible, including allocated costs. If you are not responsible for a cost and it is in your budget, eliminate it. If there is something not in your budget but you have responsibility for it, include it in your budget. Research and Development must be fully accountable for all of its responsibilities. Although you may want to break out and identify extraordinary items for accounting on some major projects, try and keep the system simple or you will spend more time on accounting than on product development.

With the exception of the costs for personnel, facilities, maintenance and comfort factors, there is no better way to address and control direct product development costs than by having a clear, measurable, and attainable objective and sound project management practices. Having a sound Research and Development strategic plan should serve as a template for cost management. Rather than always thinking in a "cost cutting" mode, think about using what some might think of as nontypical outside resources to enhance your assets. Involvement in consortiums and universities, use of consultants and contract laboratories and customer partnering are but a few examples. Make sure you understand where your funding is going in consortiums and universities. Make sure you have an active role in setting program direction.

It is becoming all too common that the major expense in new product development is not the cost of developing the chemistry and engineering, but the associated regulatory and toxicology costs. These latter two elements are increasingly becoming the drivers as to what product development a company will do or if it will do any at all. Typically, toxicology and regulatory issues go hand in hand. Many times they are combined as "regulatory expenses," the expression used here. Keep in mind that some countries and states are less interested in the efficacy and hazards of a product than they are in collecting a fee, tariff, or tax.

For alterations to an existing chemical structure, regulatory costs are about $30,000. Regulatory costs for a new biocide active ingredient can be in excess of $1 million (in U.S. cost alone). In addition, 23 other countries require new chemical notification. Nine additional counties will soon require it. China and Mexico understand the value of hard U.S. currency and require import notification. Standard notification in Australia is $8,700. The cost of testing to support a new substance notification in Canada is about $185,000 (Canadian), and test requirements are the same as in Europe.

As vendors, our cost for new product development, whether based on an incremental change in a molecule or a major new product, is typically small compared with that of our customers in the household and I&I markets. A household products company may include the following items in its product development cost through one year of national sales: R&D staff allocation, R&D expenses and capital; market research, advertising and promotion and test marketing; manufacturing capital, scale-up and plant trials; human and animal testing; and state and federal registrations. Associated costs for a new household product introduced into an existing category may approach $25,000,000. For an "improved" version, costs may be half that amount. The difference between an EPA-registered and nonregistered product is about $1.5–2,000,000. The cost for developing corresponding I&I products is about 10% that of a household product but can run as high as 20%.

Value

The second element of the value equation is return (value = return/investment). This is typically viewed as a number which is generated by marketing and for which marketing is accountable. Although the responsibility and accountability will continue to lie with marketing, the increased use of interdisciplinary strategic and tactical teams results in more shared responsibility for the generation of this number. The resultant number is more likely to be an accurate reflection of the expected return.

One would be hard pressed to read an annual report of any corporation today and not encounter the term "shareholder value." This is continually emphasized as is "customer value." R&D must recognize itself as a key provider of measurable value to both the shareholder and the customer. The authors of *Third Generation R&D* emphasize the partnership among corporate, business, and R&D management, and the cooperation of these elements to establish overall R&D strategies that are tightly linked with business and corporate strategies. As pointed out by the authors, any disconnect between any of the functions decreases the probability of success and reduces the value to shareholders and customers.

The chemical industry (and our segments, in particular), is very competitive and this competition forces greater demands on R&D. To maintain the state-of-the-art R&D resource required to give us a commercial advantage, we must be able to demonstrate, through measurement, the connection between R&D and the business and, thereby, the value of R&D. The successful companies of this decade and beyond will be those who excel in the core competency of new product development and the commercialization of new products. This is the life blood of the company.

The old paradigm of research and new product development is rapidly changing for vendors. For many consumer product and I&I companies, it has already changed. There is no question that those who do not change—for whatever reason—will be the modern-day dinosaurs of our industry. The drivers are change, flexibility, and speed. Thus, it is necessary to examine your process for new product development from top to bottom. The process is as important as the product. You may want to include things in the process that you once thought did not belong there. Too often we think the process is only from product concept to manufacture, with a tactical focus. To avoid that pitfall, begin to look at total strategy development, value measurements, more customer partnering, executive portfolio reviews, and communication outside of R&D. Once that is accomplished, it must be continually improved. Your future and the future of your company depend on it.

References

1. Roussel, P.A., K.N. Saad, and T.J. Erickson, *Third Generation R&D: Managing the Link to Corporate Strategy,* Harvard Business School Press, Boston, 1991.
2. Industrial Research Institute, *Research Technology Management,* Vol. 38, 1995.

Chapter 13

Specialty Chemicals for Washing PET Beverage Bottles

James Lichorat

DiverseyLever, Plymouth, Michigan

Introduction

In the I&I (Industrial & Institutional) Chemical Specialties Marketplace, a rebirth of an old, neglected product range has occurred: bottle wash products. Soft drink (carbonated) beverage bottle washing began over 60 years ago with the introduction of mass-marketed soft drinks in glass bottles. These were sold full of soda pop and returned to the bottler for cleaning and refilling. The "cleaning" was accomplished in huge automatic bottle washing machines, using soak and spray methods. Very important to our business were the cleaning chemicals: caustic soda built with polyphosphates, gluconates, and a range of surfactants for foam control, detergency, and rinsability.

But over the past 25 years, the glass soft drink bottle has virtually vanished in America, replaced first by steel cans, then aluminum cans, and then plastic bottles. All of these new containers are used once and then discarded (and maybe recycled). But they were not cleaned; thus, bottle wash machines and bottle wash chemicals became rare in the U.S. Product development for new bottle wash specialty chemicals became even more rare. But when two very large international soft drink companies introduced a new refillable plastic bottle 7 years ago, and hence introduced a washable bottle, new bottle wash products were suddenly very much required. The development cycle of a major new product line, which has succeeded in meeting the requirements of this new plastic bottle, is outlined here.

The Bottle

The returnable, washable, refillable bottle is made of polyethylene terephthalate (PET), just as is the more familiar "one-way" PET bottle used in the U.S. However, the refillable PET bottle is a much thicker-walled bottle, and it weighs significantly more. This washable bottle is variously termed REF PET, RPB, RET PET, or R-PET. Although it may contain fivefold more resin by weight than the one-way nonrefillable bottle, the R-PET bottle is designed to provide about 20 "trips" or "fill/wash/refill" cycles. Hence, less PET per se is ultimately required.

R-PET is used in Europe, Latin America, Asia, and Africa, but not in the U.S. or Canada. In the U.S. and Canada, bottle washing machines in soft drink bottling halls were discarded long ago, and to reintroduce these machines would be very costly. Similarly, to reintroduce a handling/distribution mechanism for R-PET would be very difficult. Because the glass returnable soft drink bottle still thrived outside the U.S., bottle wash machines, and the associated business and transportation systems still existed—allowing an introduction of R-PET to these places with minimum logistical disruption.

Coca Cola™ introduced the R-PET bottle in Germany in 1987, with Pepsi Cola™ launching at about the same time in Holland. The marketing reasons for these ventures are not the subject of this presentation. Suffice to say it was a combination of upsizing bottles to move more product, cola wars in general, and environmental and regulatory forces that allowed R-PET to become the success story it is.

Although R-PET is a true success, the previously used phrase "an introduction of R-PET to these places with *minimum logistical disruption*" is somewhat false. Part of the "disruption" occurred right in our laboratories, as we found ourselves faced with the challenge of washing (and sanitizing) a new food-beverage container. As bottlers soon discovered, the use of old bottle wash products and procedures originally developed for glass bottles was not suitable for R-PET.

The Specialty Chemical Development Stage

The first step was to assess our new challenge from both logistical and technological angles. The former meant recognizing that we needed to work very closely with the cola companies' corporate and regional divisions, and with their technical departments in the U.S., while simultaneously supporting their associated beverage bottling companies in Europe, Asia, Africa, and Latin America. In addition, we soon found it necessary to work closely with both PET resin suppliers and PET bottle producers, and ancillary vendors such as adhesive suppliers. The approach to this aspect of the challenge was a series of alignments with each segment. Perhaps the most significant was the alignment with the cola companies themselves, as we brought our own corporate technical resources directly to their corporate technical resources, to begin a technical dialogue on all aspects of R-PET bottle handling. (There are several other product application areas in addition to bottle washing in which our Packaging Hall products come in contact with R-PET bottles: conveyor chain lubrication, bottle warmers, etc.). The value of these customer alignments was proven as we developed not just new bottle wash products, but also new bottle wash machine controllers, and ultimately a new bottle wash program.

A fundamental realization was the critical role of application expertise in achieving optimized R-PET bottle washing. That is, the method of using chemicals, time, and temperature in the automated bottle wash machine emerged as the final critical factor in achieving success. A baseline factor in achieving R-PET bottle wash success is the development of "proper" or "optimized" chemical detergents. Achieving this type of product was largely a laboratory-driven exercise, whereas developing the application expertise was an "in-field" exercise.

Why are new products required for washing R-PET? Why not use those glass bottle wash products developed many years ago? Glass bottles are usually washed at about 80°C. R-PET bottles deform severely at that temperature and are washed at

60°C. Hence, some of the detergency (and sanitation) effects of higher temperatures are sacrificed and must be replaced by enhanced chemical performance. PET bottles are subject to stress cracking induced by the chemical environment of a bottle washer. In particular, the caustic soda detergent solutions can lead to stress cracking of the bottles (usually at the base ring). Higher concentrations of caustic soda lead to significantly higher stress-cracking problems. Typically, glass bottles are washed with caustic concentrations of 1.5–3.0%, but when a particular detergency problem occurs, such as mold removal problems, it is always possible to increase both the caustic concentrations and temperatures in the bottle washer to aggressively attack and solve detergency problems. With R-PET, no such aggression is possible: the bottles would be damaged beyond use. The level of caustic must be kept as low as possible, and, very importantly, the actual concentration of caustic soda in these washers must never be allowed to spike upwards, because even short contact time with high concentrations can have damaging effects. This presents a challenge to automated control because fine caustic concentration control is critical to R-PET bottle damage minimalization.

Next to the caustic concentration itself, the surfactants used in the built caustic detergent product have the most significant stress-cracking impact. We cannot predict with certainty which surfactants will be most or least damaging. Because of new test methods for measuring R-PET stress-cracking tendencies, we can screen surfactants for their suitability in R-PET bottle washing. Generally, anionic surfactants are acceptable, but because foam is anathema to these bottle wash machines, the high-foaming anionics must be avoided—or defoamed. The behavior of nonionic surfactants used for detergency, rinsing, and foam control is more difficult to predict with respect to stress cracking of PET in 1.5–3% NaOH solutions. There is a need for very low-foam nonionic surfactants for the R-PET bottle wash market.

Current test methodologies for accurately measuring R-PET stress cracking are not simple. They require an investment in resources, both capital and human. The best test procedures still require soaking/immersing whole R-PET bottles in wash solutions, rinsing, and then filling with carbonation (the pressurization is critical to producing the type of stress cracking that actually occurs). In addition, the pressurized bottles must be stored in temperature- and humidity-controlled cabinets or rooms.

As for detergency testing, simple visual tests, combined with extensive microbiology testing is the norm. Bottles are tested to be sure sanitation standards are met. Generally, detergency problems are not significant with R-PET, although detergent additives have been developed with both detergency/sanitation and stress-crack laboratory tests as in-house validation.

The Application Expertise Development Stage

After in-house validation, the extensive (and expensive) field validation stage begins. Sometimes products are actually tested and validated by the beverage companies themselves. In addition to that, third-party testing laboratories may be required for product validation before the product can be tested in the field, in a real bottle wash machine. All of these "validations" are basically "approvals" at various levels: interim, regional, restricted, and nonrestricted. Obtaining "approvals" for a single product can take 6–18 months. Fortunately, it is during these extensive field evaluations that the critical knowledge required to manage a whole R-PET bottle wash program is developed, as development chemists, microbiologists, and engineers learn the nuances of R-PET bottle wash machines and bottle wash machine and product performance.

While this overview has focused on the bottle washer caustic wash sections, the R-PET bottle is actually subjected to one or two recirculated pre-rinses, two or more soak and spray caustic wash sections, and two or more recirculated rinse sections, before undergoing a final potable (and sometimes "treated") rinse as it exits the bottle washer some 35 minutes after entering this railcar-sized bottle wash machine. The bottle washer itself is intimidating because it takes in over a thousand bottles per minute and hides all of its working parts well within its closed framework. As one learns how the machine itself operates, one can then bring that critical knowledge back into the product development laboratory to be used in the next development project. Without this intimate application expertise, the development cycle would not be successful.

Future Development Needs

The refillable plastic beverage market is fast changing. A polyethylene napthalate (PEN) bottle has already been launched in South America. We are currently working with PEN bottles in our laboratory and in Uruguay to ensure that R-PEN is successful. Perhaps new products will be required. Information gathered from the introduction of new, highly sophisticated, computerized, laser-beamed all surface empty bottle inspectors (ASEBI) suggests that we may require a new generation of rinse aids for cleaned plastic bottles. It seems that the beaded water droplets found on clean plastic surfaces are read as "contaminants" by the ASEBI. On clean glass, the classic wetted surface does not have this "beading," and hence there is no misreading by ASEBI. It may become necessary to change the plastic surface itself or to alter the rinse water's wetting/sheeting properties. No doubt, as we maintain close customer alignment, we will continue to find new development opportunities for refillable plastic bottle products. It has been a fascinating journey, around the world (literally), following our customers' use of newly placed and newly developed technologies.

Chapter 14

Eco-Efficiency: Industry's Path to Sustainability

Kenneth Alston

    S.C. Johnson & Son, Inc., Racine, Wisconsin

Introduction

Today, we all deal with one of the most important challenges of our time—the development of business practices which improve the state of the Earth's economy as well as the environment. Just three years ago in 1992, the largest gathering ever of world leaders took place in Rio de Janeiro at the so-called "Earth Summit" to discuss global environmental issues. In the time leading up to the Rio Summit, a coalition of 48 business leaders from around the world established the Business Council for Sustainable Development (BCSD). This group advised national leaders attending the World Summit about opportunities for sustainable development. Last year, BCSD merged with the World Industry Council for the Environment (WICE) to form the World Business Council for Sustainable Development (WBCSD). Their charter is to encourage businesses to integrate sustainability into their operations.

    Sustainable development encompasses meeting the needs of the present without compromising the ability of future generations to meet their own needs. Increasingly, we all recognize that economic gains cannot be achieved by using the Earth's natural resources as if there were no cost. Clearly, economic growth, social progress, and sound management of environmental resources are inextricably linked to the future health of the planet. Industry has a tangible impact on the environment. The reality of processing materials into finished products and services has had some negative impact on the environment. Because industry has contributed to the problem, it has a responsibility to take the initiative in identifying and implementing environmental solutions.

Eco-Efficiency

Eco-efficiency is industry's contribution to sustainable development. First coined by the BCSD in its "Changing Course" report to the Rio Summit, eco-efficiency is simply defined as using fewer and more efficient materials to create more, while reducing waste overall. This "use less, waste less" approach can almost always be equated with "costs less" in the long term. It encompasses two approaches: pollution prevention and pollution reduction. It links the front-end integrated product design process with manufacturing and distribution and, importantly, encompasses use and disposal to take a life-cycle approach.

    S.C. Johnson's Chairman, Sam Johnson, is a founding member of the BCSD, and an avid environmentalist and successful businessman. As a businessman, he hates waste because it is something purchased that was not sold. By integrating eco-efficiency within businesses, care of the environment becomes a tangible business imperative, not simply a "feel-good" factor. At S.C. Johnson, the corporate culture is heavily influenced by the fact that the business is family-owned and privately held. The long-standing corporate philosophies of the 108-year-old company were formalized in 1976 in a document entitled, "This We Believe." This document, providing definition to S.C. Johnson's specific family values, included a stated commitment to respecting and protecting the environment in its products and operations.

Background of S.C. Johnson, Inc.

Acting upon family environmental values has been evident throughout the company's history. As far back as 1935, H.F. Johnson, then president, traveled to Brazil to gain firsthand knowledge of the ecological impact of harvesting carnauba palm leaves for an ingredient used in our waxes. He found it to be sustainably harvested. Some have called this the first environmental audit. In the 1950s and 60s, the company pioneered water-based formulation technology in various household product categories. Later in 1975, the then president and current chairman, Sam Johnson, made a solitary, voluntary decision to cease the use of CFC propellants in our aerosols worldwide. From the mid-1980s on, our manufacturing operations around the world began to upgrade their environmental management capabilities. At the time, we called it a "cost savings" measure, because by reducing waste, we reduced waste handling costs. Today we see this decision as eco-efficient, or an economically and environmentally "win-win" situation.

S.C. Johnson believes in responsible environmental management for two simple reasons: It is the right thing to do, and it is good business. The company's philosophy is this: that where you can, you should leave things as they are; where you cannot, you should tread very lightly. We have benefited from this philosophy in many ways: it has increased our efficiency, sharply reduced emissions to the environment, and provided international recognition for our efforts in pollution reduction and prevention.

Corporate Guidelines

If you consider the function of the corporate office, it has *four* primary roles. The specific responsibilities of the corporate staff focus on the following:

- Policy (leads in establishing global policy for Safety, Health, and Environment).
- Objectives, strategies, and key action programs.
- Government relations.
- Communications: establishes global policy, principles, guidelines for all audiences, including the management of all corporate environmental and safety communications.

In 1990, when the corporate office was established, we asked, and answered, three fundamental questions. The answers to these had to address the corporate objective established by the senior management of the company: essentially, to be among the leaders in responsible environmental management, a decision central to ensuring long-term success in the global marketplace.

1. What do we want to be?   Environmental Policy Statement
2. Where do we want to go?   Global Long-Term Goals
3. How are we going to get there?   Annual Plans

All of this work was based on the premise that "What gets measured, gets done." Looking back at our benchmark in 1990, we have made significant progress.

We have phased out of over 98% of the chemicals which we identified as no longer desirable and reduced the use of volatile organic chemicals as a ratio to formula by over 14%. We have reduced the use of virgin packaging as a ratio to formula by 22.3% through minimization programs and incorporation of increased recycled content, thereby moving ahead of our ultimate 1995 20% reduction goal. We are on target to achieve the 1995 goals for manufacturing emissions during 1995, with our combined air emissions, water effluents, and solid waste as a ratio to production down 45.1% in 1994.

Leadership requires two things: substantive improvement and communication. Fundamental to our communications is that people understand that no matter what we do as a company—either in products or processes—we will have an environmental impact; what we have put into action is sustained *progress* towards minimizing our impact everywhere we operate. This "progress vs. perfection" message remains central to our vision and communication. It is central to maintaining our integrity and credibility with all of our stakeholders—which include our business partners: customers, government, and environmental community.

Eco-Efficiency as a Goal

From an industrial standpoint, waste has always been a sign of inefficiency. In the past, the major issue was how waste affected costs. Today, we know that waste bears not only a dollar value but also a quality of life value because it may be adversely affecting the environment. We also appreciate that the waste we do deal with is a resource. If it is not a scarce resource *now*, it will be in the future. Consequently, eco-efficiency is a critically important industry goal and should be the standard for every industry sector. Succeeding, however, means carrying out a parallel effort of both prevention and reduction.

Increasingly, members of industry are recognizing the business benefit of eco-efficiency and are leading with this strategy. The 3M company has its well-known 3P program "Pollution Prevention Pays" which has now moved into a new "3P-Plus" next generation program. Other large firms such as Dow, Ciba-Geigy, Xerox, and Polaroid are following suit. The challenge for formulators and packaging professionals today is knowing how to pick the projects that will have the greatest impact. We have to concentrate on finding smarter and finer trade-offs between business and environmental concerns, acknowledging that, in almost all cases, it is impossible to obtain something for nothing. Not only are we searching for the big win-wins—for the environment and the bottom line—but we are also protecting shareholder value by finding ways to improve our long-term environmental efficiency.

Eco-Efficiency Dilemmas

By integrating eco-efficiency into business decision making in this way, we make a positive contribution to both environmental and financial improvements—a true win-win scenario. It sounds simple, but in reality we also face some difficult dilemmas. Some dilemmas are played out in the media. Loggers or owls. Ranchers or wolves. Others, such as those we face, may be less newsworthy, but still pose difficult questions. Paper or plastic. Natural or synthetic. Formulation and packaging choices are no longer as simple as they once seemed. In the recent past, corporate responsibility was straightforward—maximize capital gains and dividends and do not become entangled in social responsibility; manage stockholders' money but do not presume to manage their social conscience. What business was that of corporate management? Well, it is our business. We are not above the values of society. Companies do not have to make a choice between being socially and environmentally responsible and meeting shareholders' financial objectives. That choice is a myth, a red herring, a false dilemma. It is an excuse for inaction—or worse.

At S.C. Johnson, our commitment to environmental values does not mean that we do not face dilemmas when it comes to making these environmental values real. We do. Here is a typical environmental choice that a company seeking to do the right thing will face. At S.C. Johnson, marketing is of central importance. And what is the dilemma between marketing and the environment? In marketing, you want a large sign board. You want your packaging to be big enough to command shelf space and to stop people. You want your product to be the best on the market and competition today is fierce. When we introduced an air freshener called Glade Plug-Ins™, competitive pressures caused our marketing people to want to put this compact product in a relatively large package. Our environmental ethic causes us to desire smaller, less wasteful packaging. We eventually chose the smaller, but that is not always an easy choice to make in a competitive market. As another example, consider the product called Future, which is a clear floor polish. The original bottle was virgin plastic and was absolutely clear—a great showcase for the crystal clear product. The new bottle is recycled plastic. It is not quite as strong as the previous bottle, and in addition, it has a brownish tone. We are marketing this clear product, for a clear shine, and what are we doing? We are putting it in a brownish bottle. This runs counter to basic marketing principles, but it *has* resulted in an annual savings of 638,000 pounds of virgin material.

Challenges for the Future

The challenges facing S.C. Johnson and industry in general are clear: we must develop quality products that meet customers' needs and wants, and manage our processes in ways that minimally affect the environment. As we move forward, we must face up to a frustrating reality that despite all of the success stories, many businesses are at the first stage in their environmental position; they do not yet accept eco-efficiency as a real business benefit, and they choose not to participate in the debate regarding future environmental progress. That must change if we are to have any possibility for success of a sustainable world. We all have to stop viewing the environment as all cost and no gain, and start recognizing that waste and pollution are measures of inefficiency.

President Clinton's Council for Sustainable Development (PCSD) is finalizing its report on two years of deliberations on sustainable development policy options. Eco-efficiency is a key theme in the report to the President. We expect that there will be a growing recognition that we share responsibility for the life-cycle environmental burden of our products and services within the whole supply chain. It really is up to each of us as individuals to play our part and make eco-efficient decisions as we make choices in our personal and professional lives. If we, as representatives of industry, succeed in this regard, eco-efficiency will indeed be industry's next achievement. It will be our path towards a more sustainable future and we will bequeath a much healthier planet to our children.

Chapter 15

The Need for Multifunctional Surfactants

Arshad Malik, Ned Rockwell, and Y.K. Rao

Stepan Company, Northfield, Illinois

Background

Surfactants are known for their ability to alter the physical properties of solutions, particularly at surfaces and interfaces. The effect of surfactants on solution behavior is typically measured by physicochemical properties such as surface- and interfacial tension reduction or cloud/clear point determination (1). Evaluation is also typically performed using application methodologies such as fiber wetting, foaming behavior, cleaning efficiency, and soil mitigation (2).

However, the term "multifunctional surfactant" implies that certain surfactants are capable of providing attributes which are not commonly associated with surfactants. For example, surfactants are not typically associated with hydrotropy. Hydrotropy is commonly described as the ability of a specific compound (i.e., a hydrotrope) to increase the solubility of relatively insoluble organic compounds. This phenomenon is particularly useful, for example, if a product designer intends to create a homogeneous liquid product with long shelf-life and needs to avoid the use of other materials such as solvents. One mechanism of hydrotropy is the disruption of the relatively stable (and prone to phase separation) surfactant liquid crystalline phase by migration and incorporation of a hydrotrope into the micelle (Fig. 15.1) (3). The structure of one class of compounds known to be hydrotropic is well known: a short-chain hydrophobe with a polar hydrophilic head group. This structure of the hydrotrope acts as a phase coupler in mixed water/oil systems through the mechanism mentioned above (4).

Lamellar Liquid Crystalline Phase of Surfactant

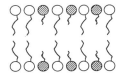

Unstable Liquid Crystalline Structures in Presence of Hydrotrope

Fig. 15.1. Mechanism of hydrotrope action.

The structure of a typical surfactant is similar to that of a hydrotrope but differs by having a relatively long hydrocarbon chain. This long chain is much less water soluble than the short chain of these hydrotropes and this prevents most surfactants from having much hydrotropic character. As will be shown, this generalization is not true for every surfactant. The results of some of the following investigations demonstrate that some compounds exhibit behavior not normally associated with their particular class of material. The job remains for the formulator to capitalize on the uniqueness of these compounds to satisfy today's demanding product requirements.

Materials

A broad range of compounds was selected for this study. Table 15.1 provides a summary of materials and the abbreviations used for each material. Our intention was to select a range of materials across the entire surfactant/hydrotrope spectrum. The materials included in this study were selected as a function of their common use, their related structures (alkyl groups joined asymmetrically to polar head groups), and relatively unique behavior. Obviously, it is impractical to attempt to display all of the materials in this format.

Methods

Surface tension. A Krüss K-12 Tensiometer was used to determine the surface tension of solutions at 25°C. The Wilhelmy plate method was used to measure the surface tension in dynes/cm. Deionized water was used in all experiments.

TABLE 15.1 Materials Used

Sodium xylene sulfonate	SXS
Sodium octane sulfonate	SAS-C8
Sodium hexane sulfonate	SAS-C6
Sodium octyl sulfate	SOS
Sodium decyl sulfate	SDS
α-Sulfo methyl laurate, sodium salt	ASMEL
α-Sulfo methyl stripped cocoate, sodium salt	ASMEC
Decyl diphenyl oxide disulfonate	DOD-C10
Dodecyl (branched) diphenyl oxide disulfonate	DOD-C12
Octyl dimethyl amine oxide	AO-C8
Decyl dimethyl amine oxide	AO-C10
Phosphate ester [C8/C10 alcohol ethoxylate (5 mol EO)]	PE
Nonyl phenol ethoxylate (9 mol EO)	NPE
Ethoxylated amide	

Draves Wetting. The Draves wetting test [American Society of Testing Materials (ASTM) Designation D-12281, (5-g cotton skein, 3 g hook)] measures the ability of surfactants to cause solutions to wet or spread evenly onto surfaces. The surfactant with the shortest wetting time generally works faster. Test conditions were as follows: temperature, 25°C; concentration, 1.0% and 0.1%; deionized water.

Ross-Miles Foam. The Ross-Miles foam was tested according to ASTM Designation D-1173-53 under the following conditions: temperature, 25°C; concentration, 1.0% active; deionized water.

Cloud Point. Effect of additive on cloud point of 1% solution of C12–C15 linear alcohol ethoxylate (7 mol EO). The percentage of additive to be tested in total surfactant = (amount of additive ÷ C12–C15 linear alcohol ethoxylate) × 100.

Hard-Surface Cleaning. Tests were done according to Modified ASTM Designation D-4488-91, Section A-5 under the following conditions:

- Substrate: white vinyl floor tiles
- Soil: urban oily/particulate soil
- Test apparatus: Gardner straight-line washability apparatus
- Reflectance instrument: Hunter Lab (spectrophotometer)
- Detergent use level: 0.6% active to 12% TKPP
- Water hardness: 140 ppm
- Temperature: 25°C
- Evaluation: percentage of soil removal calculated

Laundry. *Terg-O-Tometer* testing was used. Surfactants were evaluated at 0.1% active concentration in 140 ppm hard water. Three cotton and polyester/cotton (65/35) fabric swatches soiled with dust-sebum represented the laundry load. *Terg-O-Tometer* conditions were as follows:

- Wash cycle   10 min
- Wash temperature   38°C
- Rinse cycle   5 min
- Rinse temperature   26°C
- Agitation speed   100 rpm

Detergency was measured as an increase in reflectance between the soiled and cleaned swatches (5).

Bleach Stability. The base formula was as follows: surfactant, 2.0% active; sodium hypochlorite (NaOCl), 4.0% active; NaOH (50%), 1.0%; NaCl, 0.25%; water, QS to 100. For stability results, the base formulation was heated to 86–90°C for 4 h and then tested for bleach remaining. Control sample (no surfactant) result was designated to be 100%.

Results and Discussion

Surface Activity, Wetting and Foaming

The multifunctional surfactants were more surface active than the conventional hydrotrope, sodium xylene sulfonate (SXS). The data shown in Table 15.2 clearly indicate that all of the materials used in this study are very surface active at 1% active concentration, reducing the surface tension of water to low values. An industry workhorse, SXS, at 1% reduced the solution surface tension to 58 dynes/cm, while at 0.1% concentration, the surface tension of solution remains virtually unchanged from that of pure water, 72.4 dynes/cm. This suggests that SXS is a relatively poor adsorber at the interface. This is supported by the fact that the Draves wetting time for SXS is more than 10 min compared with less than a minute for several multifunctional ingredients at 1.0% active concentration (Table 15.3). Figure 15.2 shows the Ross-Miles foam profiles for the experimental materials. It is clear that SXS has insignificant foaming properties (zero foam at 5 min) suggesting again that it does not absorb well at the interface. The multifunctional ingredients showed moderately good foaming behavior. While SXS did not show good surface-active, wetting, or foaming properties, several multifunctional agents, specifically sulfo methyl esters, phosphate esters, and amine oxides, were significantly better.

Hydrotrope Coupling

Hydrotropic efficiency was evaluated by determining the amount of hydrotrope (SXS) or multifunctional agent required to clarify 100 g of three different detergent base for-

TABLE 15.2 Effect of Various Surfactants on the Surface Tension of Deionized Water (72.4 dyne/cm at 25°C)

Ingredient	0.1% Active	1.0% Active
	(dyne/cm)	
AO-C10	27	27
SDS	29	29
PE/ASMEL/ASMEC/DOD-C12	33	32
DOD-C10	39	37
SAS-C8	43	26
SAS-C6/SOS	48	28
AO-C8	53	32
SXS	72	58

TABLE 15.3 Draves Wetting

Ingredient	0.1% Active	1.0% Active
ASMEC	45 s	27 s
PE	127 s	2 s
SDS/AO-C10	>10 min	2/3 s
SAS-C8	>10 min	10/13 s
ASMEL/SAS-C6	>10 min	23/27 s
DOD-C12/AO-C8/DOD-C10	>10 min	134/222/390 s
SXS	>10 min	>10 min

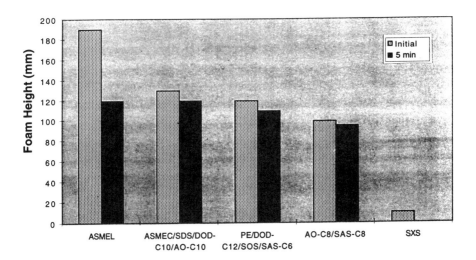

Fig. 15.2. Ross-Miles foam profiles (1.0% active, deionized water, 25°C).

mulations: an ethoxylated amide base, a high hydrophilic lipophilic balance (HLB) nonionic base, and a mixed high/low HLB nonionic base. The results obtained from this study, as well as the compositions of the base formulations, are shown in Tables 15.4, 15.5, and 15.6. The results indicate that the multifunctional ingredients require significantly less usage than SXS to clarify any given formulation. The sharpest contrast between the performance of multifunctional agents and SXS occurs in the high/low HLB nonionic base formulation.

Cloud Point Elevation

The multifunctional ingredients were very effective in increasing the cloud point of 1% solution of C12–C15 linear primary alcohol ethoxylate (7 mol EO), even at concentrations as low as 1% (Fig. 15.3). Sulfo methyl esters, phosphate ester, SDS, etc.

TABLE 15.4 Hydrotrope Coupling Performance—Ethoxylated Amide Base[a]

(% active of hydrotrope required to clarify 100 g of base formulation at 50°C) Ingredient	Weight %
ASMEC	2.8
ASMEL/AO-C8	3.2
SAS-C6	3.6
SOS/SAS-C8	4.3
PE/AO-C10/DOD-C12/DOD-C10/SXS	5.0
SDS	5.5

[a]*Formulation:* ethoxylated amide, 10.0%; sodium metasilicate • $5H_2O$, 10.0%; tetrapotassium pyrophosphate (TKPP), 5.0%; deionized water, Balance.

TABLE 15.5 Hydrotrope Coupling Performance—High Hydrophilic Lipophilic Balance (HLB) Nonionic Base[a]

(% active of hydrotrope required to clarify 100 g of base formulation at 60°C) Ingredient	Weight %
AO-C8	1.6
AO-C10/PE/DOD C10	1.9
SAS-C8	2.2
SAS-C6/SOS	2.5
ASMEL	3.0
ASMEC/DOD C12	3.4
SXS/SDS	4.1/4.6

[a]*Formulation:* nonyl phenol ethoxylate (9 mol EO), 5.0%; tetrapotassium pyrophosphate (TKPP), 10.0%; deionized water, Balance.

TABLE 15.6 Hydrotrope Coupling—High Hydrophilic Lipophilic Balance (HLB)/Low HLB Nonionic Base[a]

(% active of hydrotrope required to clarify 100 g of base formulation at 25°C) Ingredient	Weight %
DOD-C10	0.5
AO C10/SDS/ASMEL	1.0/1.1/1.2
SOS/DOD-C12	1.3/1.4
ASMEC/AO-C10	1.5/1.6
PE/SAS-C8	1.9
SAS-C6	2.2
SXS	4.4

[a]*Formulation:* sodium citrate, trisodium, 1.0%; nonyl phenol ethoxylate (9 mol EO), 5.0%; nonyl phenol ethoxylate (4 mol EO), 5.0%; deionized water, Balance.

increased the cloud point by at least 25°F, whereas SXS at an additive concentration of 4% had almost no effect. In addition, C8 amine oxide behaved similarly to SXS, whereas C10 amine oxide increased the cloud point by 10°F.

Hard-Surface Cleaning and Laundry

Performance evaluation studies were conducted for all of the materials under standardized hard-surface cleaning and laundry conditions in the laboratory. Figures 15.4 and 15.5 show the data in graphical form for hard-surface cleaning and detergency performance, respectively. All multifunctional ingredients performed better than SXS in these studies. The percentage of cleaning efficiency by the Gardner soil removal method was highest for a 9.5 mol nonylphenol ethoxylate and is significantly higher for sodium decyl sulfate (SDS) and C10 amine oxide than for SXS.

Sulfonated methyl esters (ASMEC) performed extremely well in soil removal for all types of fabrics by Terg-O-Tometer detergency testing as seen in Fig. 15.5 and was comparable to alcohol ethoxylates, whereas SXS, short-chain sulfonates, amine oxides, etc. performed very poorly in the detergency tests.

Bleach Stability

SXS had very good bleach stability (90% of control base formulation, Fig. 15.6). With the exception of sulfo methyl esters (ASMEC, ASMEL) and phosphate ester, all other ingredients had very good bleach stability.

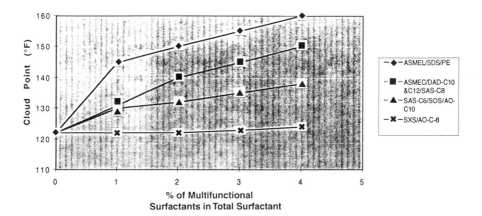

Fig. 15.3. Effect of multifunctional surfactant on cloud point (1% solution of C12–C15, 7EO linear primary alcohol ethoxylate).

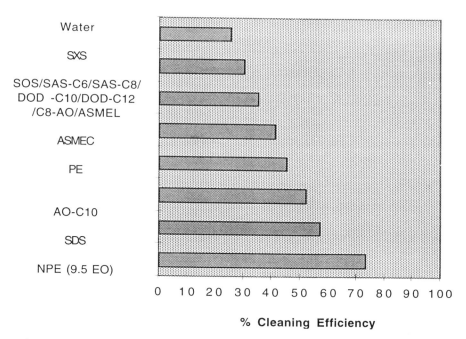

Fig. 15.4. Hard surface cleaning (Gardner method)—0.6% surfactant 0.12% TKPP. For abbreviations, see Table 15.1.

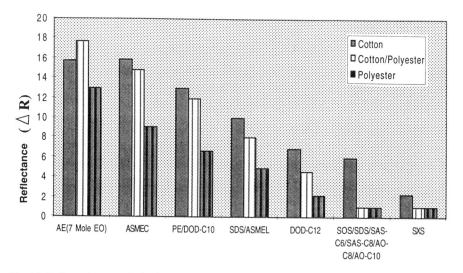

Fig. 15.5. Laundry test cloth cleaning performance.

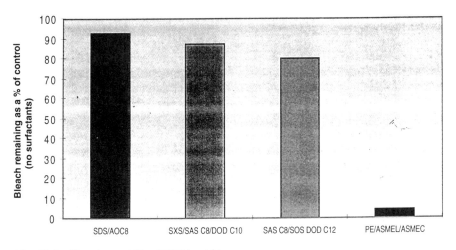

Fig. 15.6. Cleaning stability (90°C for 4 h).

Conclusions

These results clearly demonstrate that formulators now have a wide choice of multifunctional materials that offer several advantages over traditional materials such as SXS. Multifunctional materials, in addition to being equal or better hydrotropes than SXS, exhibit significantly better surface activity, wetting, foaming and cleaning characteristics. Specifically, sulfo methyl esters, amine oxides, short-chain sulfates and sulfonates exhibit properties significantly superior to those of SXS. On the other hand, SXS has a long track record as a good hydrotrope with good bleach stability.

Development of new multifunctional surfactants will continue to expand the ability of formulators to customize and therefore differentiate their products. Thus, it is important for formulators to be cognizant of the specialized functions various potential ingredients may perform and be constantly on the lookout for new materials which will create more desirable attributes for their products in the most efficient manner.

Acknowledgments

The authors wish to thank Liz Hampe for providing publishing assistance and Henry Watanabe for his technical assistance.

References

1. Rosen, M.J., in *Surfactants and Interfacial Phenomena,* John Wiley & Sons, New York, 1988.
2. Shah, D.O., *Chemical Engineer Education 14* (1977).
3. Friberg, S.E., and J. Chiu, *Dispersion Sci. & Technol. 9:*443 (1988).
4. Ward, A.J.I., C. Marie, L. Sylvia, and M.A. Phillip, *J. Dispersion Sci. & Technol. 9:*149 (1988).
5. Rockwell, N., and Y.K. Rao, in *Proceedings of World Lauric Oils Conference,* Manila, edited by T. Applewhite, AOCS Press, Champaign, 1994, p. 138.

Chapter 16

Overview of European Regulatory Activities

Pierre V. Costa

FIFE/AIS, Brussels, Belgium

Introduction

A considerable amount of legislation affecting the soap, detergent and maintenance products industries in Europe is based on political and economical dimensions. Today a wide range of products are regulated at the European level which, when correctly implemented in the different European countries, has the advantage of replacing divergent national regulations by one single Community Act. The European Union (EU) is not simply a treaty between sovereign states. It has a common internal border inside which EU legislation applies exactly like domestic legislation. To adopt a common Community Act, the EU works with important institutions; the European Commission with headquarters in Brussels proposes the EU legislation voted by the Council and the Parliament. The Union covers a market of some 362 million consumers. It is a market with constant growth in gross domestic product (GDP). Ideally, this is only the beginning: the long-term goal is to bring all democracies of Europe together, including countries of Central and Eastern Europe. Some of these countries have already signed association agreements (Poland, Czech Republic, Rumania, Slovakia, Bulgaria). Free trade pacts with the three Baltic States were also concluded last year. EU legislation, which takes priority over domestic legislation, covers an increasing range of issues affecting the soap, detergent, and maintenance products industries.

Product-Related Legislation

Detergent and maintenance products are covered by regulations on chemicals. First, the pair of directives, the Dangerous Substances Directive dealing with chemical substances and their classification, packaging and labeling, and the Dangerous Preparations Directive dealing with preparations and products made up of chemical substances. The Dangerous Substances Directive was enacted in 1967 and the Dangerous Preparations Directive was issued 21 years later. The older directive, concerning substances, has been amended seven times and has been subject to 20 so-called adaptations to technical progress. The Dangerous Preparations Directive has also been adapted three times to technical progress, and its first amendment Directive is presently under discussion. Such frequent changes are not easy to deal with.

These two directives have also had a large number of offshoots, including a directive concerned with child-resistant packagings, tactile warnings of danger, and safety data sheets. This might be considered unimportant to the detergent and maintenance products industries because most of the cleaning products are not dangerous. This is largely true, although some "borderline" consumer products and a number of institutional and industrial products are so classified. For example, some machine dishwashing products are classified as "irritant." But every product is a chemical preparation. Thus, considerable work is involved in examining all products to determine how they would be classified if classification should ultimately be required.

Second, there is the Draft Biocidal Products Directive, which was originally called the Non-Agricultural Pesticides Directive. The Directive was drafted with good intentions. There is an existing Directive to deal with agricultural pesticides, which is generally approved of. A number of pesticides are used outside of agriculture, thus it seems reasonable to have a similar Directive to deal with nonagricultural uses of pesticides, e.g., products such as rat poison or wood preservatives. Unfortunately, the scope of this directive as drafted is very wide.

The definition of biocidal products in the draft is as follows:

> active substances and preparations containing one or more active substances, put up in the form in which they are supplied to the user, intended to destroy, deter, render harmless, prevent the action of, or otherwise exert a controlling effect on any harmful organism.

This definition is so broad that it could be considered to include many detergent cleaning products.

The Existing Substances Regulation requires a dossier with the effects of all substances on the market, starting with large tonnage substances and gradually moving down the scale. This includes substances that are sold as constituents of other products. This is a massive burden to industry in terms of cost and man hours spent simply in collecting the information in a very short period. The aim of the 1993 legislation is to evaluate the risks of existing substances to man and the environment and to recommend appropriate control measures and risk reduction strategies, where necessary. The evaluation is preceded by data collection of all substances listed in the European Inventory of Existing Commercial Chemical Substances (EINECS), which are produced or imported in quantities above 10 metric tons per annum. The data are collected in three phases depending on the tonnage. Priority chemicals are regularly selected by applying criteria laid down in the legislation. The risk of the substances appearing on the priority lists is subsequently assessed following agreed-upon principles.

There is a third set of directives, the Nominal Quantities/Unit Pricing Directives. These Directives are also a pair of linked Directives. In 1980, the Nominal Quantities Directive laid down for a variety of products a set of quantities (weight and capacities) in which the products are sold which must be allowed free access to all European markets. Again, this was an appropriate piece of legislation produced in 1980 at that stage of the development of the European Internal Market. In 1988, the Unit Pricing Directive was brought in, which required unit pricing of a large number of goods. This was to apply to products that were not sold in these nominal quantities and was to be enacted as of June 1997.

Again this seems sensible. Unit pricing allows consumers to compare value between different products in different weights. Unfortunately, the sizes in the Nominal Quantities Directive have never been updated. Technical changes and other developments have meant that, in our industry, many of the most common sizes are not on the list. So far as washing powders are concerned, the old-style detergents are on the list but the newer compacts and refills are not. The latter two were put on the market to reduce environmental impact. This would have the interesting result that the majority of the products in the market would have to be unit priced, but the small minority in these old-fashioned weights would not. Industry has been pressing for many years for these weights to be updated.

Further, there is the Ecolabeling Regulation establishing a European environmental labeling system so that products with a better environmental performance can be labeled as such by an official scientific system which would be a common system in Europe rather than competing national systems. Laundry detergents were identified as a priority category. Criteria were developed over the last four years. Products eligible for ecolabeling must pass the wash and wear test. Ecolabels will be granted for only three years, and an updating process will be continued. There are eight ecological criteria parameters that apply to ingredients and one packaging criterion. Criteria were published in the European Union Official Journal in autumn of 1995, enabling companies to apply for an EU ecolabel.

Toilet soaps are covered by the Cosmetics Directive 76/768 which contains positive lists (ingredients which can be used) and negative lists (ingredients which must not be used). There have been six amendments to the Cosmetics Directives and about 18 adaptations to technical progress.

The Directive on Misleading Advertising of 1984 is presently under discussion. The directive grants the consumer the benefit of a reversal of the burden of proof, by enabling courts to request evidence from an advertiser of the truth of his factual claims, insofar as this is sensible. It also provides that member States ensure that adequate and effective means exist for the control of misleading advertising. Such means include legal provisions under which persons or organizations regarded as having a legitimate interest in prohibiting misleading advertising may take legal action against such advertising or bring such advertising before an administrative competent authority.

Environmental Legislation

EU legislation in the field of the environment consists of approximately two hundred pieces of legislation covering a wide range of sectors, including management of waste and chemicals, biotechnology, water, and air pollution. Furthermore, several "horizontal" measures relating to the environmental impact assessment of certain public and private projects and access to environmental information have been adopted.

EU legislation on the environmental control of industrial installations includes Directives on pollution caused by certain dangerous substances in the aquatic environment of the EU and its offshoot Directives, on combating air pollution from industrial plants, and on major accident hazards of certain industrial activities. An important proposal on integrated pollution prevention and control, which will deal with all forms of pollution from stationary sources, is presently being discussed within the EU Council. In addition, the Council adopted in 1993 a regulation establishing a voluntary eco-audit scheme enabling companies to have their environmental performance certified.

In relation to water pollution, and in addition to those Directives relating to discharge from stationary sources or pollution from specific industries, a number of legal acts laid down quality objectives or other requirements of water intended for specified uses: water for human consumption and bathing. Thus, the revised Drinking Water Directive includes boron as a substance to be controlled. The central purpose of this Directive is to ensure that water for human consumption is safe. The latest draft shows a sevenfold reduction in the level allowed which could well have a substantial effect on sodium perborate included as a bleach in many household detergents. The Directive is the type of Community legislation which has the widest and most constant relevance to the soap, detergent, and cleaning products industry. It is by Directives that the harmonization process among members of the EU has been undertaken.

The EU Treaty also gives industry the opportunity to make voluntary agreements which can be incorporated as EEC Recommendations published in the EEC Official Journal. Such a Recommendation on ingredient labeling exists for cleaning products. It is a pan-European industry voluntary agreement incorporated in a 1989 EC Commission Recommendation. This complicated system could be avoided.

Conclusions

Many of the regulations are very detailed and prescriptive, rather than imposing general duties. Many amendments and technical adaptations are required. Also, achieving acceptable compromises among the 15 member States, which offer a mosaic of economic and political interests, cultural patterns and moods, is a very tricky problem. It will become even more complex in the 21st century when countries from Central Europe join the Union. Equal treatment will help to bridge the gap between Western and Eastern economies. These events will bring new elements to the developments of the EU and change its priorities, its way of operation, and its structure. In most of the Eastern countries, we helped to set up national associations of soaps and detergents which are now in the process of becoming members of AIS and FIFE. The enlarged European Union is committed to coordinated or even joint action. EU legislation shows no sign of slowing, and with the accession of new member States, the future shape of Europe will be different. In particular, rules governing the protection of the environment will undoubtedly be stricter.

The formation of the European Union has been good for industry—and the operation of the European market is much improved in just a few years. Changes in Europe have helped economic integration. We in industry recognize the rationale for legislation when it is based on real needs in helping to achieve a single market across Europe. We are committed as an industry to continue environmental improvements.

However, to avoid problems that put progress at risk, we require the following:

- Better implementation and policing in all countries which have adopted EU Acts
- Legislation based on the best available science and knowledge
- Realistic cost-benefit analysis done before legislation is introduced
- A less prescriptive and detailed approach to legislation
- A more measured pace to legislation

These modifications will benefit both industry and the environment.

Chapter 17

Generation 2000 Appliances

John T. Weizeorick

Association of Home Appliance Manufacturers, Chicago, Illinois

Introduction

A discussion of futuristic home laundry appliances relies on the presence of a well-defined set of goals for energy efficiency as well as some preliminary ideas on how to achieve these goals. The present political situation in Washington, D.C. has resulted in some dispute in the industry. Energy requirements for the next generation of home laundry appliances remain unresolved. This has muddied the waters in terms of future product design. The home laundry members of the Association of Home Appliance Manufacturers (AHAM) are working to keep abreast of any new decisions.

Assuming that Washington will not discontinue the appliance energy program, it is worthwhile to consider likely regulations and their impact. At this time we believe that the home laundry appliances introduced at the end of this decade will be new designs which are efficient and use less water than today's appliances in response to government regulations on energy and water conservation; they will likely require changes in the formulation of detergents and rinse additives and, as a consequence, they will cause both the machine and detergent industries some risk because this new generation of product will be designed to meet governmental and environmental mandates rather than consumer wants and needs.

Background

The U.S. appliance industry is subject to a regulatory scheme set up by the National Appliance Energy Conservation Act of 1987 (NAECA), which gives the Department of Energy (DOE) the authority and responsibility to establish minimum energy efficiency performance standards for certain kinds of household equipment. NAECA has set standards for products since 1990. Products failing to meet these performance standards could not be sold in the U.S. The DOE is also authorized to periodically update these standards. The first update for home laundry appliances occurred with the establishment of a minimum efficiency standard for clothes washers and clothes dryers effective May 14, 1994.

The 1994 clothes washers are substantially more efficient than the previous generation of products, resulting in a 21% increase in energy factor over the 1993 products (Fig. 17.1). The energy factor, which is measured in terms of cubic feet of basket volume per kilowatt hour per cycle, went from 1.00 in 1993 to 1.21 in 1994. The actual energy consumption per cycle of the 1994 units was reduced almost 18% over 1993 models: from 2.71 to 2.23 kWh/cycle. These figures are shipment weighted averages. Consumption includes both the electrical energy consumed by motors and other electrical components and the energy used to heat the water.

Figure 17.2 indicates the proportion of energy used for U.S. household appliances including home laundry. The residential sector represents about 20% of the total U.S. end use consumption of energy and, within that 20%, clothes washers account for about 4.4% and clothes dryers about 4.1%. The efficiency improvement attained by 1994 clothes washers to meet the government mandated standard was achieved largely by tinkering with the present designs, that is, major product redesigns were not necessary to achieve the required level of efficiency. Motors and other electrical components were upgraded, hot and cold water mixes in the various cycles were modified, and other fine tuning was done, but the basic electromechanical design we have grown familiar with since the introduction of automatic washers was maintained.

Changes on the Horizon

The engineering required to meet 1994 DOE standards was considered "fine tuning" rather than evolutionary change. As we move towards Generation 2000 Appliances, the rate of change is likely to accelerate and become more evolutionary. The design "drivers" are changing, and the home laundry industry is under pressure from environmental groups as well as electric utilities and regulatory commissions to reduce both energy and water consumption. The successful commercialization of alternate tech-

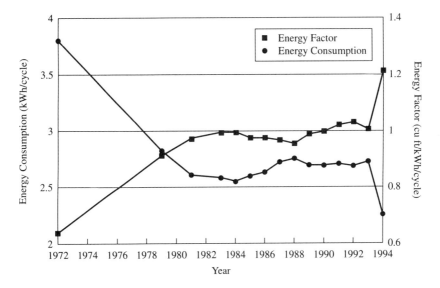

Fig. 17.1. Energy factor and energy consumption in clothes washers.

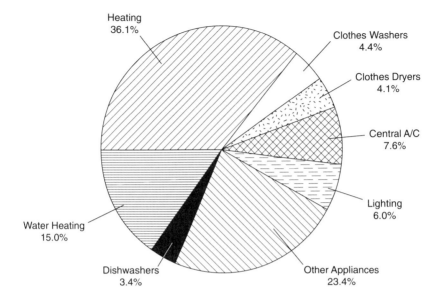

Fig. 17.2. U.S. household energy consumption from all energy sources; 1993 data.

nologies in other parts of the world which approach the goals of these groups is used to demonstrate technical feasibility. The U.S. market generally considers these technologies unsuitable—but they may be mandated by government requirements.

Alternate Technology

An example of an alternate technology is the tumble-type washer which is sometimes called the horizontal axis or front-loading washer (Fig. 17.3). This technology represents about 75% of the European market but less than 1% of U.S. sales. U.S. consumers prefer top-loading machines but a survey done by A.D. Little for AHAM in 1991 found U.S. consumers reluctant to accept the top-loading horizontal axes (tumbler) washer. U.S. manufacturers are reluctant to switch entirely to such designs because of the high cost of retooling and revamping manufacturing facilities (estimated for the industry to be in the area of $1.4 billion), as well as the uncertainty that the U.S. consumer would accept such products. For comparison purposes, the net worth of the home laundry industry in the U.S. is about $933 million.

Test Procedure—Washer and Dryer

The DOE will ultimately decide the issue by prescribing an energy standard. The energy consumed will be measured by a standard test method also prescribed by the DOE. The home laundry industry and the DOE agree that the existing clothes washer test procedure, which was developed in the 1970s, is outdated and design restrictive

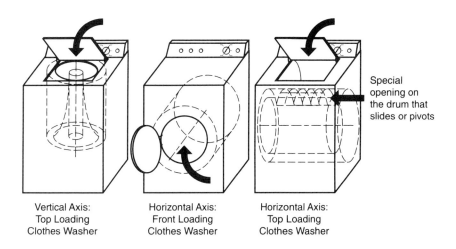

Fig. 17.3. Different types of automatic clothes washers.

and must be modified. One of the issues to be addressed is that clothes washing alone does not constitute home laundry. A significant portion of the loads washed in an automatic washer are dried in an automatic dryer and thus home laundry energy consumption includes energy use by both appliances. Significant amounts of energy savings are available in drying if clothes enter the dryer with lower initial moisture level than available today. This can generally be achieved by spinning in the clothes washer at a higher speed and/or for a longer time.

Modified and Future Energy Factor

The home laundry industry has already submitted its proposal for a modified test procedure for clothes washers to the DOE. It includes a method for calculating a modified energy factor based on the final moisture retention provided by the clothes washer. The present DOE energy standard requires standard size washers to meet or exceed an energy factor of 1.18; the environmentalists have stated that they want Generation 2000 washers to achieve an energy factor on the order of 3.2. This is a 60% reduction in energy used for washing clothes. While some of this energy may be saved in the dryer through the use of higher washer spin speeds, the rest will have to achieved by other means.

Conclusions

We can speculate that some manufacturers will choose the tumble-type washer approach. In fact, some have already announced that they will introduce tumble-type washers in the next several years. Others are likely to look at other innovative technologies. One manufacturer which originally stated it was looking at the tumble-type approach has since advised that it is abandoning that technology for another approach which it believes will be more acceptable to the U.S. market. In addition, the trade press has reported on other innovative designs which have been developed in other countries, including the air-bubble washing machine developed in Korea which claims high efficiency.

Detergent manufacturers and their suppliers must assimilate all of this information and make the business decisions necessary to assure that they have chemical products available for use in the Generation 2000 Appliances. Because the future risks are high, they must be managed rather than passively accepted. Each company should have a working relationship with home laundry manufacturers to anticipate the new designs that they are developing so that environmentally safe, performance-enhancing chemical products that meet the needs of Generation 2000 washers can be developed. We know that some of those will be tumble-type, but we can also look forward to other innovative means for providing the mechanical energy for the clothes washing process.

To meet the 1994 energy standard, manufacturers reduced the quantity and temperature of the hot water used. It is likely that there will be further reduction of the thermal energy available for washing. Greater chemical energy may be necessary for future machines that use less water at cooler temperatures. The new clothes washing machines may also introduce greater turbidity and a wash system that is significantly different than present day products. This assumes of course that the existing political situation in Washington does not permanently discontinue the appliance energy program for the foreseeable future.

It is important that our industries strengthen our partnership to be certain that our mutual customers, the consumers of the U.S., are not disadvantaged by the new household laundry energy standards that are under consideration.

Index

A

ADDs (automatic dishwashing detergents). *See* Dishwashing detergents
AES. *See* Alcohol ether sulfate (AES)
AHAM. *See* Association of Home Appliance Manufacturers (AHAM)
Air Resources Board. *See* California Air Resources Board (ARB)
AIS. *See* Association Internationale de la Savonnerie et de la Detergence (AIS)
Alcohol ether sulfate (AES), 18–19
Alcohol ethoxylate, 18–19
Alkyl polyglycosides, 19, 21
Alkylphenol ethoxylates, 15
All surface empty bottle inspectors (ASEBI), 133
Aluminosilicate zeolites. *See* Zeolites
American Society for Testing and Materials (ASTM), 5, 8
Aminocarboxylate use, 97–100
Amorphous sodium silicates. *See* Silicate use
Amylases, 61, 114–115
Anhydrous sodium silicates, 26
Antimicrobial compounds, 5, 10–11, 20
classification as pesticides, 10–11
proposed regulation of, 6. *See also* Biocidal Products Directive, Draft
Antiredeposition aids (ARDs), 46
Appliances. *See* Generation 2000 appliances
Aquifer contamination, 86
ARB. *See* California Air Resources Board (ARB)
ARDs. *See* Antiredeposition aids (ARDs)
ASEBI. *See* All surface empty bottle inspectors (ASEBI)
Assessment of cleaning products and processes, 132–133, 140–142
life cycle, 2, 8
in vitro *versus* in vivo, 11–12, 15
Association Internationale de la Savonnerie et de la Detergence (AIS), 13, 152
Association of Home Appliance Manufacturers (AHAM), 154, 156
ASTM. *See* American Society for Testing and Materials (ASTM)
Automatic dishwashing detergents (ADDs). *See* Dishwashing detergents

B

Bar soaps. *See* Personal washing bars
BCSD. *See* Business Council for Sustainable Development (BCSD)
Bentonite, 23
Biocidal Products Directive, Draft, 150
Biodegradability, 49–51, 91
boundary, 105
defining, 50
of enzymes, 558
Bleaching
using enzymes, 61
Blending
dry, 38–39
high-shear, 38–40
Boron in cleaning products, 13
Bottle wash chemicals. *See* Refillable polyethylene terephthalate (R-PET) bottle
Builder materials, 23–24, 31–34, 36, 47–48, 106–107, 111
precipitating, 53
sequestering, 53
Business Council for Sustainable Development (BCSD), 134

C

Calcium carbonate precipitation, 43–44
Calcium sequestration, 26, 31
California Air Resources Board (ARB), 3–4
Capillary fringe contamination, 86
Carboxylates, 33. *See also* Polycarboxylates
Carcinogenic agents, 14
Cellulases, 61, 116–117
Cellulose degradation, mechanism of, 117
Chelating agents, 97–112
Chemical Manufacturers Association (CMA), 11

Chemical Use Inventory (CUI), 17
Chemophobia, 2
Chromatography, head-space, 82
Clean Air Act, 4
Clinton, President William, 138
   Executive Orders of, 6
Cloud point, 141, 143–145
CMA. *See* Chemical Manufacturers Association (CMA)
CMC. *See* Critical micelle concentration (CMC)
Cocoyl N-methyl glucose amides, 19, 21
Cogranules, 25, 34
Cola wars, 131
Complex stability, 103
Conditional stability constant, 99–103
Consumers, 1
   changing habits of, 20–21
   educating, 1, 8–10
   skepticism of, 2
Continuous processing, 39–40
Cosmetic, Toiletry, and Fragrance Association (CTFA), 5
Cosmetics Directive, 151
Critical micelle concentration (CMC), 64–66, 71, 73, 76, 79–80, 89
Critics of cleaning product industry, 2
CTFA. *See* Cosmetic, Toiletry, and Fragrance Association (CTFA)
CUI. *See* Chemical Use Inventory (CUI)

D

Dangerous Preparations Directive, 149
Dangerous Substances Directive, 149
Dense non-aqueous phase liquids (DNAPLs), 79, 86, 88–89
Department of Energy (DOE), 54, 154–157
Departments of Natural Resources (DNRs)
   educational efforts by, 9
Departments of Water Resources (DWRs), 16
Detergents. *See* Laundry detergents
Dialkyl diphenylether disulfonate (DADS), 70, 75–77
Dialkyltetralin impurities, 18

Diethylene triamine pentaacetate (DTPA), 97, 101
Diphenylether (DPE) sulfonates, 75
Diquaternary ammonium compounds, 71
Directive on Misleading Advertising, 151
Dishwashing detergents, 23, 98
   for hand use, 19–20
   incorporating enzymes, 58–59
   nonphosphate, 53–54, 106, 110
δ-disilicate, 27–28, 31, 36
Disinfectant products. *See* Antimicrobial compounds
DNAPLs. *See* Dense, non-aqueous phase liquids (DNAPLs)
DNRs. *See* Departments of Natural Resources (DNRs)
Dodecyl pyrridinium bromide (DPBr), 64–67
Dodecyl pyrridinium chloride (DPC), 68
DOE. *See* Department of Energy (DOE)
DPBr. *See* Dodecyl pyrridinium bromide (DPBr)
DPC. *See* Dodecyl pyrridinium chloride (DPC)
DPE. *See* Diphenylether (DPE) sulfonates
Draize eye irritancy test, 11–12
Draves skein wetting times. *See* Wetting times (WOT), Draves skein
Drinking Water Directive, 152
Dual-compartment packaging, 40
Dye transfer inhibitors, 48–50

E

Earth Summit, 134
ECETOC, European Center for Ecotoxicology and Toxicology of Chemicals (ECETOC)
Eco-auditing, 152
Eco-efficiency of products, 8, 134–138
Ecolabeling Regulation, 151
Ecology *versus* cost and performance (ECP) diagrams, 31–34
EDTA. *See* Ethylene diamine tetraacetate (EDTA)
EDTMP. *See* Ethylene diamine tetramethylene phosphonate (EDTMP)

Education of consumers. *See* Consumer education
EH&S. *See* Environmental health and safety (EH&S) impacts
EINECS. *See* European Inventory of Existing Chemical Substances (EINECS) list
Encapsulation technologies, 40–41
End users. *See* Consumers
Endocrine disruptors, 14–15
Energy factor, 154–155
Enhanced oil recovery (EOR) technology, 85, 87–88
Enthalpy, 64, 66
Environmental attributes of products, 134–138
  of surfactants, 79–96
  voluntary disclosure of, 7–8
Environmental health and safety (EH&S) impacts, 8
Environmental Protection Agency (EPA), 3–4, 7–10, 13–17
  Report to Congress, 4
Enzyme activity, evaluating, 118–124
Enzyme coding, 59
Enzyme use
  background of, 113–114
  in detergent products, 57–62, 113–125
  future of, 60–61
  in nondetergent products, 14
EOR. *See* Enhanced oil recovery (EOR) technology
EPA. *See* Environmental Protection Agency (EPA)
Estrogenic agents. *See* Endocrine disruptors
Ethanol
  disposal of, 3
Ethylene diamine tetraacetate (EDTA), 97–99, 101–107
Ethylene diamine tetramethylene phosphonate (EDTMP), 97, 107
European Center for Ecotoxicology and Toxicology of Chemicals (ECETOC), 13
European Inventory of Existing Chemical Substances (EINECS) list, 108, 150

European regulatory activities, 149–153
European Union Official Journal, 151–152
Existing Substances Regulation, 150
Eye irritation potential, 11–12

F

Fabric incrustation, 43–45, 47–49, 106–107
Fabric softeners, 18–19
FDA. *See* Food and Drug Administration (FDA)
"Federal Acquisition, Recycling, and Waste Prevention" (Executive Order), 6
Federal Implementation Plans, 3
Federal Insecticide, Fungicide and Rodenticide Act (FIFRA), 10–11
Federal Supply Service (FSS), 7–8
Federal Trade Commission (FTC), 7, 13
FIFRA. *See* Federal Insecticide, Fungicide and Rodenticide Act (FIFRA)
Fillers, 29
Filming, 53–54
Foam fractionation, 84
Food and Drug Administration (FDA), 4–5, 91
  Tentative Final Monograph (TFM), 4–5
Formulated products, 37
  prejudices against, 2
Formulators of products, 1
FSS. *See* Federal Supply Service (FSS)
FTC. *See* Federal Trade Commission (FTC)

G

Gemini surfactants. *See* Surfactant strategies
General Services Administration (GSA), 7–8
Generation 2000 appliances, 154–158
Genetic engineering, 59, 62
German Chemical Substances Act, 108
Gibbs free energy, 64–66
Global impacts. *See* Impacts, scale of, global *versus* local
Government agencies, 1–2
Greywater use, 15–16
GSA. *See* General Services Administration (GSA)

"Guides for the Use of Environmental Marketing Claims," 13

H

Hard water. *See* Water
Hard-surface cleaners, 19–20, 141, 145–146
HDLs. *See* Laundry detergents, heavy-duty liquid detergents (HDLs)
Health-Care Antiseptic Drug Products. *See* Over-the-counter (OTC) Health-Care Antiseptic Drug Products
Healthcare Continuum Model, 5
Heavy metal ion removal. *See* Chelating agents; Metal ion removal
Heavy metals in cleaning products, 12–13
Hectorite, 23
Household hazardous waste (HHW) disposal, 9
Hydrolysis mechanisms
  of peptide bonds, 118–120
  of starch, 115
  of triglyceride, 116
Hydrophilicity, 51
Hydrophobicity, 63–69
  of polyester, 52
Hydrotropicity, 74–77, 139–140, 142–144
Hydroxycarboxylate use, 97–99

I

I&I. *See* Industrial & Institutional (I&I) Chemical Specialities Marketplace
Impacts
  reversibility of, 7
  scale of, global *versus* local, 7
Incrustation. *See* Fabric incrustation
Industrial & Institutional (I&I) Chemical Specialities Marketplace, 127–128, 130
Industrial Research Institute, 126
Information, ensuring accuracy of, 9
Ingredients, increasing numbers of, 1
Interface affinity, 60, 73, 91
International Sanitary Supply Association (ISSA), 11
International Union of Biochemistry (IUB), 114, 118
Inventory Update Rule (IUR). *See* Toxic Substances Control Act (TSCA)
ISSA. *See* International Sanitary Supply Association (ISSA)
IUR. *See* Toxic Substances Control Act (TSCA)

K

Kanemite, 26
Kirk-Othmer Encyclopedia of Chemical Technology, 39
Krafft point boundary, 19, 90

L

LAS. *See* Linear alkylbenzene sulfonate (LAS)
Laser examination of plastic bottles. *See* All surface empty bottle inspectors (ASEBI)
Laundry detergents, 18–19, 23. *See also* Builder materials
  color-safe, 48–49
  complex liquid products, 39–40
  granular, 18–19
  heavy-duty liquid detergents (HDLs), 18–19, 107, 112
  high density powders, 37–39, 47, 58
  incorporating incompatible ingredients, 40–41
  manufacturing history of, 37–41
  superconcentrated, 30
  temperatures used at, 20–21
  trends in, 34–41
Life cycle assessment. *See* Assessment of cleaning products and processes
Light non-aqueous phase liquids (LNAPLs), 86–87
Linear alkylbenzene sulfonate (LAS), 18–19, 22, 75
Lipases, 60–61, 115–116
LNAPLs. *See* Light non-aqueous phase liquids (LNAPLs)
Logistical disruption, minimizing, 131

M

MADS. *See* Monoalkylated disulfonate (MADS)
Makatite, 26
MAMS. *See* Monoalkylated monosulfonate (MAMS)
Masking metal ions, 97
Media influence, 1
Metal ion removal, 80–82. *See also* Chelating agents
Metasilicates, crystalline. *See* Sodium tripolyphosphate (STPP)
Methyl ester sulfonates, 18
Methylglycine diacetic acid (MGDA), 105–112
MEUF. *See* Micellar-enhanced ultrafiltration (MEUF)
MGDA. *See* Methylglycine diacetic acid (MGDA)
Micellar-enhanced ultrafiltration (MEUF), 79–85
Micellization, 63–68, 70–71
 entropy of, 63
Monoalkylated disulfonate (MADS), 75
Monoalkylated monosulfonate (MAMS), 75

N

NAECA. *See* National Appliance Energy Conservation Act (NAECA)
NAPLs. *See* Non-aqueous phase liquids (NAPLs)
National Appliance Energy Conservation Act (NAECA), 154
Nitrilo triacetate (NTA), 97–99, 105–107
Nominal Quantities/Unit Pricing Directives, 150–151
Non-agricultural pesticides, regulating. *See* Federal Insecticide, Fungicide and Rodenticide Act (FIFRA)
Non-aqueous phase liquids (NAPLs), 86
NTA. *See* Nitrilo triacetate (NTA)

O

Organization for Economic Cooperation and Development (OECD), 15
screening test, 105, 107
Over-the-counter (OTC) Health-Care Antiseptic Drug Products, 4

P

3P program, 136
p(Asp). *See* Polyaspartate [p(Asp)]
Patent trends, 19, 30–31
PBS. *See* Public Buildings Service (PBS)
PCAs. *See* Polycarboxylates (PCAs)
PCSD. *See* President's Council for Sustainable Development (PCSD)
PEN. *See* Polyethylene napthalate (PEN)
Pentasodium tripolyphosphate (STP), 97, 99
p(EO/VA). *See* Poly(ethylene oxide/vinyl acetate) [p(EO/VA)]
Personal washing bars, 41, 151
PET. *See* Polyethylene terephthalate (PET)
Phosphate use, 12, 22, 30, 33, 97
 as sequastrant for metal ions, 97–98
Phosphonate use, 97–99
Phosphorous and Nitrogen Removal from Municipal Wastewater: Principles and Practices, 12
Point of sale consumer decisions
 certifications appropriate to, 6
Polyaspartate [p(Asp)], 33
Polycarboxylates (PCAs), 23, 43–44, 47–49
 polymeric, 33
 production efficiencies with, 46
Polyethylene napthalate (PEN), 133
Poly(ethylene oxide/vinyl acetate) [p(EO/VA)], 52
Polyethylene terephthalate (PET), 51–52
Polymer-based technologies, 42–56
 flexibility of, 42
 future of, 46–55
 history of, 43–46
Polyphosphate use. *See* Phosphate use
Polysoaps, 83
Polyvinylpyrrolidone (PVP), 48–50
Practice for Environmentally Preferable Cleaners/Degreasers (ASTM draft), 8
President's Council for Sustainable Development (PCSD), 7, 138

President's Regulatory Reinvention Initiative, 13
Product claims, 7
Product containers. *See* Residue standards
Product introduction costs, 128
Professional organizations, 1
Professional strength products, inappropriate use of, 5
Proteases, 117, 119, 122
Protein engineering. *See* Genetic engineering
Public Buildings Service (PBS), 7
Public interest groups, 1
PVP. *See* Polyvinylpyrrolidone (PVP)

R

Raw materials suppliers, 1
Refillable polyethylene napthalate (R-PEN) bottle, 133
Refillable polyethylene terephthalate (R-PET) bottle. *See also* Polyethylene terephthalate (PET)
  washability of, 130–133
Regulatory costs, 127
Regulatory groups. *See* European regulatory activities; Government agencies
Release mechanisms, 40
Renewability, 22
Research and development
  costs, 126–128
  return on investment, 128–129
Research Technology Management, 126
Reserve alkalinity, 29
Residue standards, 11
Risk assessment procedures, 2, 5
Ross-Miles foam, 141–143
R-PEN. *See* Refillable polyethylene napthalate (R-PEN) bottle
R-PET. *See* Refillable polyethylene terephthalate (R-PET) bottle

S

Safe Drinking Water Act, 13
Safety concerns
  groups involved, 1–17

human *versus* environmental, 2
Safety data sheets, 149
SARs. *See* Structural activity relationships (SARs) of chemicals
Scale inhibition, 110
SCAS. *See* Semi-continuous activated sludge (SCAS)
SDA. *See* Soap and Detergent Association, The (SDA)
SDS. *See* Sodium dodecyl sulfate (SDS)
SEAR. *See* Surfactant-enhanced aquifer remediation (SEAR)
SED. *See* Semiequilibrium dialysis (SED)
Semi-continuous activated sludge (SCAS), 50
Semiequilibrium dialysis (SED), 82–83
Septic tank systems, 15
Silicate use, 23–36, 43
  amorphous, 25, 32
  insoluble, 23
  layered, 23–36, 47
  soluble, 23
Silicic acids, 26
Silicon dioxide, 26
Silicon hydroxide, 26
Silicon monoxide, 26
SIPs. *See* State Implementation Plans (SIPs)
Skin irritation factor, 73
Soap and Detergent Association, The (SDA), 1–2
Soda ash, 23, 32
Sodium carbonate, 25, 43
Sodium citrate, 33, 53
Sodium dodecyl sulfate (SDS), 90
Sodium polyacrylate, 45, 50, 54
Sodium polyaspartate, 50–51
Sodium silicates. *See* Silicate use
Sodium sulfate, 29, 32
Sodium tripolyphosphate (STPP), 23, 25, 33, 43–45
Sodium xylene sulfonate (SXS), 142–147
Soil, removing pollutants from, 85–91
Soil dispersion, 45–46
Soil release polymers (SRPs), 51–52
Solubilization. *See also* Hydrotropicity
  isotherms, obtaining, 82

micellar, 89
of organic pollutants, 79–82
Spray drying process, history of, 37–38
SRPs. *See* Soil release polymers (SRPs)
Stability, bleach, 141–142, 145
Stability constant. *See* Conditional stability constant
State Implementation Plans (SIPs), 3
Stepwise dissociation, 99
STP. *See* Pentasodium tripolyphosphate (STP)
STPP. *See* Sodium tripolyphosphate (STPP)
Strategic planning, 127, 136
Stress cracking, 132
Strippability, 19, 84–85
Structural activity relationships (SARs) of chemicals, 15
Surface activity, 71–73, 140, 142
equilibrium *versus* dynamic, 76
Surfactant adsorption. *See* Hydrophobicity
Surfactant-enhanced aquifer remediation (SEAR), 87–91
Surfactants. *See also* individual surfactants by name
allowable effluent concentrations of, 83
for environmental remediation, 79–96
future requirements, 21
gemini, 70–78
having direct food additive status, 91
multifunctional, 139–148
nonionic, 19, 21
in sediments, 16
short-chain, 20
strategies, 18
SXS. *See* Sodium xylene sulfonate (SXS)

T

TCE. *See* Trichloroethylene (TCE)
Temporal aspects of impacts. *See* Impacts, reversibility of
Tentative Final Monograph (TFM). *See* Food and Drug Administration (FDA)
Teratogenic agents, 14
Terg-O-Tometer testing, 141, 145
Ternary diagrams. *See* Ecology *versus* cost and performance (ECP) diagrams
Testing. *See* Assessment of cleaning products and processes
TFM. *See* Food and Drug Administration (FDA)
Third Generation R&D, 126, 128
Toxic Substances Control Act (TSCA), 17, 108
Toxics Release Inventory (TRI). *See* Toxic Substances Control Act (TSCA)
Trends in cleaning products, 18–20, 28–30, 128–129
TRI. *See* Toxic Substances Control Act (TSCA)
Trichloroethylene (TCE), 79, 85
Tripolyphosphates. *See* Sodium tripolyphosphate (STPP)
TSCA. *See* Toxic Substances Control Act (TSCA)

U

UF. *See* Micellar-enhanced ultrafiltration (MEUF)
Ultrafiltration (UF). *See* Micellar-enhanced ultrafiltration (MEUF)

V

Vadose zone contamination, 86
Volatile organic compounds (VOC)
safety of, 2–4

W

Washing machines, standards for, 54, 154–155
Wastewater discharge, 12–13
reclaiming. *See* Greywater use
Water. *See also* Drinking Water Directive; Safe Drinking Water Act
hard, 25–26, 28, 106. *See also* Wastewater discharge
removing pollutants from, 79–85
WBCSD. *See* World Business Council for Sustainable Development (WBCSD)
Wettability, 51. *See also* Hydrotropicity

Wetting times (WOT), Draves skein, 76, 141–143
What Can I Do? (free SDA brochure), 10
WHO. *See* World Health Organization (WHO)
WICE. *See* World Industry Council for the Environment (WICE)
"Work Practices for Handling Enzymes in the Detergent Industry" (SDA publication), 14
World Business Council for Sustainable Development (WBCSD), 134
World Health Organization (WHO), 13
World Industry Council for the Environment (WICE), 134
WOT. *See* Wetting times (WOT), Draves skein

Z

Zeolites, 23–24, 35, 43, 47
 aluminosilicate, 24, 30
 MAP, 24, 30, 32, 48